Philosophy of Techn

To my daughter Lela Dusek

PHILOSOPHY OF TECHNOLOGY

AN INTRODUCTION

VAL DUSEK

Blackwell
Publishing

BLACKWELL PUBLISHING
350 Main Street, Malden, MA 02148-5020, USA
9600 Garsington Road, Oxford OX4 2DQ, UK
550 Swanston Street, Carlton, Victoria 3053, Australia

14
.D86
2006

First published 2006 by Blackwell Publishing

1   2006

Library of Congress Cataloging-in-Publication Data

Dusek, Val, 1941–
Philosophy of technology : an introduction / Val Dusek.
p. cm.
Includes bibliographical references and index.
ISBN-13: 978-1-4051-1162-1 (hc. : alk. paper)
ISBN-13: 978-1-4051-1163-8 (pbk. : alk. paper)
ISBN-10: 1-4051-1162-3 (hc. : alk. paper)
ISBN-10: 1-4051-1163-1 (pbk. : alk. paper)
1. Technology—Philosophy.   I. Title.
T14 D86 2006
601—dc22
2005025431

A catalogue record for this title is available from the British Library.

Set in 10.5/13pt Dante
by Graphicraft Limited, Hong Kong
Printed and bound in India
by Replika Press Pvt. Ltd.

The publisher's policy is to use permanent paper from mills that operate
a sustainable forestry policy, and which has been manufactured from
pulp processed using acid-free and elementary chlorine-free practices.
Furthermore, the publisher ensures that the text paper and cover board
used have met acceptable environmental accreditation standards.

For further information on
Blackwell Publishing, visit our website:
www.blackwellpublishing.com

Contents

Introduction

As philosophy goes, philosophy of technology is a relatively young field. Courses called "History of Modern Philosophy" cover philosophers of the Renaissance and the seventeenth and the eighteenth centuries. Philosophy of the early twentieth century is covered in "Contemporary Philosophy." The main branches of philosophy go back over 2200 years. Philosophy of science was pursued, in fact if not in name, by most of the early modern philosophers in the seventeenth and eighteenth centuries. By the mid-nineteenth century several physicists and philosophers were producing works that focused solely on the philosophy of science. Only sporadically were there major philosophers who had much to say about technology, such as Bacon around 1600 and Marx in the mid-nineteenth century. Most of the "great philosophers" of this period, although they had a great deal to say about science, said little about technology. On the assumption that technology is the simple application of science, and that technology is all for the good, most philosophers thought that there was little of interest. The "action" in early modern philosophy was around the issue of scientific knowledge, not technology. The romantic tradition from the late eighteenth century was pessimistic about science and technology. Romantics emphasized their problematic and harmful aspects, and only a handful of academic philosophers concerned themselves with evaluation and critique of technology itself. Particularly in Germany, there was a pessimistic literature on the evils of modern society in general and technological society in particular. We shall examine at length several of the twentieth-century inheritors of this tradition. In the English-speaking countries, with the exception of romantic poets such as Wordsworth and mid-nineteenth-century culture critics such as Carlyle, Matthew Arnold, and Ruskin, or the socialist artist William

1

INTRODUCTION

Morris, few had much to say about the evaluation of technology. Only with Hiroshima and Nagasaki, and the realization that atom and hydrogen bombs could literally cause humanity to go extinct, did widespread, popular, critical evaluation of technology occur in the English-speaking world. With the widespread popular awareness that industrial pollution and its degradation of the environment was a major problem, perhaps dated from the publication of Rachel Carson's *Silent Spring* in 1962, or from Earth Day of 1970, a further wave of concern for the understanding of the negative side-effects of technology arose. With the advent of genetic engineering and the specter of human cloning in the late 1970s, with the possibility of technologically manipulating human heredity and even human nature, there was yet another set of issues and impulses for the critical evaluation of technology.

The Society for the Philosophy of Technology was founded in 1976, thousands of years after the birth of philosophy, over three centuries after the beginning of intensive examination of the nature of scientific knowledge, and about a century after the beginnings of systematic philosophy of science.

Not only was the philosophy of technology late in coming of age, but the field itself is hardly consolidated even now. One of the problems is that the philosophy of technology involves the intimate interaction of a number of different fields of knowledge: philosophy of science, political and social philosophy, ethics, and some aesthetics and philosophy of religion. Specialists in ethics and political philosophy have rarely been deeply involved in the philosophy of science and vice versa. Furthermore, the philosophy of technology ideally involves knowledge of science, technology, society, politics, history, and anthropology. One philosopher of technology, Jacques Ellul, even claims that since no one can master all of the relevant fields, no one can direct or deflect technology (see chapter 6).

The topics of the philosophy of technology are varied. In this book there is discussion of the relation of philosophy of science and its recent developments to the philosophy of technology (chapter 1). There is a brief discussion of the nature of definition and various proposed definitions of technology (chapter 2). The theme of technocracy, or rule by an elite of scientists and technologists, is presented in chapter 3, and also used as a means to discuss some of the historical philosophies of technology (such as those of Plato, Bacon, Marx, St Simon, and Comte). The issue of technological rationality and rationality in general is discussed in chapter 4. A variety of characterizations of and approaches to rationality are considered: formal rationality, instrumental (or means–end) rationality, economic rationality, transcendental

2

rationality, and dialectical rationality, among others. Risk/benefit analysis, a form of formal rationality, closely related to mathematical economics, and often used to evaluate technological projects, is presented and evaluated.

Next, approaches to philosophy of technology very different from the logical, formal economic, and analytical approaches are examined. Phenomenology, involving qualitative description of concrete experience, and hermeneutics, involving interpretation of texts in general, are presented in chapter 5. Several philosophers of technology who have applied phenomenology and hermeneutics to fields such as technical instrumentation and computers are discussed.

A complex of issues involving the influence of technology on society and culture are treated in chapters 6 and 7. Technological determinism, the view that technological changes cause changes in the rest of society and culture, and autonomous technology, the view that technology grows with a logic of its own out of human control, are discussed and evaluated.

Chapter 8 describes the debates concerning whether technology is what distinguishes humans from other animals, and whether language or technology is most characteristic of humans.

Chapters 9 and 10 discuss groups of people who have often been excluded from mainstream accounts of the nature and development of technology. Women, despite their use of household technology and their widespread employment in factories and in the telecommunications industry, were often omitted from general accounts of technology. These accounts often focus on the male inventors and builders of large technological projects. This is true even of some of the best and most dramatic contemporary accounts (Thompson, 2004). Women inventors, women in manufacturing, and the burden of household work are often downplayed. Similarly, non-Western technology is often shunted aside in mainstream Western surveys of technology. The contributions of the Arabs, Chinese, South Asians, and Native Americans to the development of Western technology are often ignored. The power and value of the local knowledge of non-literate, indigenous peoples of the Americas, Africa, and the South Pacific is also often ignored. However, ethno-science and technology raise issues about the role of rationality in technology and the nature of technology itself.

There is also a powerful traditional critical of technology, at least since the romantic era of the late 1700s. In contrast to the dominant beliefs about progress and the unalloyed benefits of technology, the Romantic Movement celebrated wild nature and criticized the ugliness and pollution of the industrial cities. With the growth of scientific ecology in the late nineteenth and early twentieth centuries, a new dimension of scientifically based criticism of pollution was added to the general preference for wild nature over technology. After 1970 the political ecology movement, especially in Germany and the USA, became a mass movement of criticism and even rejection of technology. Even the early nineteenth-century Luddite notion of machine smashing was revived.

Finally, in the late twentieth century the social construction of technology became a major component of the sociology and philosophy of technology. This approach opposes and criticizes technological determinism and autonomous technology theories that claimed to show that technology was following a predetermined course with a logic of its own. Instead, it is claimed that there is a great deal of contingency in the development of technology. Many interest groups have input into the final form of a given technology, and the apparent necessity of the paths of technological development is an illusion.

Because of this late beginning and the overlapping knowledge requirements of the field of philosophy of technology, it is one in which it is difficult to get initial orientation. On the one hand, there is a large, but largely superficial, literature of after-dinner speeches about "Technology and Man," written mainly by technologists and policy-makers. On the other hand, there is a highly convoluted and obscure European literature presupposing some of the most difficult and obscure philosophers (such as Hegel, Marx, and Heidegger). This continental European philosophy of technology does attempt to grapple with the place of technology in history. It is grand and ambitious, but often obscure and obtuse. Certainly the German Martin Heidegger and his students such as Arendt and Marcuse, the critical theorists of the Frankfurt school, and the French theorist of technology out of control, Jacques Ellul, are all notorious for the difficulty and obscurity of their prose. That obscurity is not solely the province of European writers is shown by the stimulating but often hopelessly muddled prose of the Canadian theorist of the media, Marshall McLuhan. Lewis Mumford, the American freelance architecture and city planning critic and theorist of technology, is readable, but at times long-winded.

Not only are major European figures (such as Heidegger, Arendt, and Ellul), whom Don Ihde has called the "grandfathers" of the field, difficult to read, there is a further complication in that many other schools of twentieth-century philosophy have contributed to the philosophy of technology. Anglo-American linguistic and analytic philosophy of science has contributed. Besides the various European schools of philosophy (neo-Marxism,

INTRODUCTION

European phenomenology and existentialism, and hermeneutics), American pragmatism and Anglo-American process philosophy have also supplied frameworks for writers on technology. (Early twentieth-century process philosophy, widely rejected by mainstream analytical philosophers, has undergone a revival among those often dubbed "postmodern" philosophers of technology, such as Bruno Latour and Paul Virilio, via the French postmodernist Deleuze. Anglo-American social constructivist sociologist of science Andrew Pickering, for instance, has become interested in Whitehead via this French revival.)

One of the goals of this book is to introduce to the beginning student the various philosophical approaches that lie behind different takes on the philosophy of technology. Analytic philosophy of science and ethics, phenomenology, existentialism, hermeneutics, process philosophy, pragmatism, and social constructivism and postmodernism are some of the philosophical approaches that we shall examine. There are chapters on phenomenology and hermeneutics, on social constructionism and actor-network theory, and on Anglo-American philosophy of science. There are also boxes inset in various chapters about aspects of the philosophy of Martin Heidegger, about process philosophy, and about postmodernism. It is hoped these will help to orient the student in a field in which such a variety of approaches and philosophical vocabularies have been deployed.

Despite the variety and difficulty of the literature of the field, there is also promise of further research in the philosophy of technology. Various often compartmentalized branches of philosophy, such as philosophy of science and political philosophy, as well as competing and often non-communicating schools of philosophy, such as those mentioned above, may become combined and synthesized through their use in the philosophy of technology.

I hope that the reader will find this work a guide to the many fascinating topics and approaches that this field encompasses.

Note: throughout the book, technical terms and names of persons that I think are particularly important for readers to know are presented in **bold** type.

1

Philosophy of Science and Technology

Much of the philosophy of technology in the nineteenth and twentieth centuries was done without consideration of or involvement with the philosophy of science. There were theoretical reasons for this, tacitly assumed by most writers. If science is simply a direct, uninterpreted description of things as they are, untainted by cultural and social biases and constraints, then science is simply a mirror of reality. Furthermore, if technology is simply applied science, and technology is, fundamentally, a good thing, then there are no special philosophical problems concerning technology itself. That is, the frameworks for the development of technology and its reception are not of interest. There are only after-the-fact ethical problems about technology's misapplication. However, recent approaches to the philosophy of science have shown that science is laden with philosophical presuppositions, and many feminists, ecologists, and other social critics of science have claimed that science also is laden with social presuppositions. Many recent approaches to philosophy of technology claim that technology is not primarily, or even is not at all, applied science.

First we shall survey the major mainstream philosophies of science from the early modern period to the mid-twentieth century, and then look at some more recent philosophies of science, and how they impact our understanding of technology.

The most widely known and accepted philosophy of science (often presented in introductory sections of science texts) has been **inductivism**. Francis Bacon (1561–1626) was not only one of the earliest advocates of the values of science for society (see chapter 3) but also the major advocate of the inductive method. According to inductivism, one starts with observations of individual cases, and uses these to predict future cases. Bacon enumerated what he called "idols," or sources of pervasive bias, both individual and social, that prevented unbiased, pure observation and logical theorizing. One of these groups of "idols," the idols of the theater, he claimed, was philosophy.

Inductivism generalizes from individual cases to laws. The more individual cases that fit a generalization, the more probable the generalization. British philosophy from the seventeenth to the twentieth centuries has been primarily inductivist. The inductivist view spread widely to other nations during the eighteenth and nineteenth centuries. By the nineteenth century the sway of inductivism was such that even philosophers who did not really follow the inductive method were claimed to have done so. The theorist of electromagnetic fields, Michael Faraday (1791–1867), is an example. In the 1800s he was widely portrayed as what Joseph Agassi (1971) has called "the Cinderella of science," a poor boy who, through careful, neutral observation, arrived inductively at major discoveries. Twentieth-century studies show that he used romantic philosophical ideas, and in his own notebooks made numerous metaphysical speculations as frameworks for his electrical conjectures (Williams, 1966; Agassi, 1971). The evolutionist Charles Darwin (1809–82) even claimed he worked "on true Baconian principles, without any theory," though his real method is one of conjecturing hypotheses and deducing their consequences (Ghiselin, 1969). Although inductivism is probably still the most widely believed account of science among the public (although not as dominant as previously), inductivism has a number of logical problems. The most fundamental is called **Hume's Problem** or the **problem of the justification of induction**. These are technical and may appear nit picking to the non-philosopher, but they are significant enough problems to cause many philosophers of science (and scientists who have thought about these issues) to move away from straightforward inductivism (see box 1.1).

Logical positivism was a philosophy that developed in central Europe in the 1930s (primarily Vienna, where the group of philosophers and scientists calling itself the "**Vienna Circle**" arose) and spread to the USA with the emigration of many of its leaders after the rise of Nazism. The logical positivists took over much of the spirit of Comte's earlier positive philosophy without its explicit social theory and quasi-religious aspect (see discussion of Comte in chapter 3). The logical positivists, like the older positivists, saw science as the highest, indeed the only, genuine form of knowledge. They saw statements other than empirically supportable ones as meaningless. This was the **verification theory of meaning**: for a statement to be meaningful it had to be possible to verify it (show it true by empirical evidence). This

Box 1.1

The problem(s) of induction

In the early eighteenth century the Scottish philosopher David Hume (1711–76) raised what is called the problem of induction. It is really the problem of the justification of induction. Hume granted that we use induction, though he thought that it turns out to be a matter of custom and habit. A later philosopher, George Santayana (1863–1952), called it "animal faith." We expect the future to be like the past. Hume raised the question of the justification or rational reason for our belief in induction. The usual answer one gives is that "induction works." Hume doesn't deny that it works. However, Hume notes that what we really mean is that induction (or science in general) "has worked in the past and therefore we expect it to work in the future." Hume pointed out that this reasoning from past success to future success is *itself* an inductive inference and it depends on the principle of induction! Thus appeal to success or "it works" is circular. It implicitly applies the principle of induction to induction itself. It attempts to use the principle of induction to justify the principle of induction. Hume showed how other attempted justifications (such as an appeal to probability rather than certainty) also fail or beg the question. Most of Hume's contemporaries didn't see the problem and dismissed Hume's claims. However, one philosopher, Immanuel Kant (1724–1804), recognized the importance of Hume's problem and called it "the scandal of philosophy" (he could have called it the scandal of science, given its implications, though most working scientists were unaware of it). Kant's solution was that principles built into the human mind, such as causality and necessity, allow us to organize our experience in ways that allow regularity and induction. The cost of Kant's solution is that the regularity of nature is no longer known in things in themselves, separate and outside of us, but is the way we structure our experience and knowledge of nature. That is, we can't know that "things in themselves" follow lawful regularities, but only that our mind is structured to seek such regularities and structure our experience in terms of such laws. Karl Popper (1902–94) accepted the insolubility of Hume's problem. However, Popper's solution involves giving up the claim that science uses induction. Thus the proposed solutions to Hume's problem lead to views of science far from the usually accepted ones. Either we structure our experience in terms of induction, but cannot know if nature really follows laws, or we do not really ever use induction, but deceive ourselves into thinking that we do.

criterion of meaning was meant to exclude theology and metaphysics from the realm of the cognitively meaningful. The logical positivists were guilty of the fallacy of persuasive definition, in that they defined "meaningless" in a technical manner, but then used the term in a pejorative, dismissive manner, equivalent to "worthless" or "garbage."

Although the logical positivists did see the spread of the scientific approach as a boon to humankind and most of them held politically reformist and often social democratic views, they did not consider political philosophy as part of genuine, "scientific" philosophy, and with a few exceptions did not explicitly discuss their broader social views within their analyses of science. The sociologist and philosopher Otto Neurath (1882–1945) was a notable exception, who explicitly referred to Marxism in a positive manner (Uebel, 1991). (Neurath also contributed to highway and airport technology by inventing the non-verbal, pictorial schemas and symbols that caution the driver about approaching curves or deer crossings, and guide the passenger or customer to the restroom today; Stadler, 1982.) However, even the implicit support of social democracy by the positivists and their American followers was suppressed during the McCarthy Era of the early 1950s in the USA (Reisch, 2005).

The "logical" part of logical positivism consisted of the reconstruction of scientific theories using formal, mathematical logic. The apparent success of Bertrand Russell (1872–1970) in reducing mathematics to logic inspired logical posivitists to systematize scientific theories in terms of assumptions (axioms) and rigorous logical deductions. For the most part they analyzed science as a body of statements or propositions. Scientific theories were treated as primarily conceptual entities. In this the logical positivists were similar to many earlier philosophers in their treatment of science. The positivists simply analyzed the structure and connection of statements with more precision and rigor than their predecessors. The inspiration of Russell's logical foundation for mathematics inspired some positivists, notably Rudolf Carnap (1891–1970), to attempt to develop a formal inductive logic. The failure of this program, carried out over decades, has convinced almost all philosophers of science that a formal logic of induction of the sort Carnap envisioned is impossible. Induction involves informal assumptions and judgment calls. (See chapter 4.)

The quest for precision and rigor led the positivists to self-criticism of their own empirical or observational criterion of meaningfulness. Though this led to the demise of the criterion (the verificationist criterion of meaning), it is a tribute to the honesty and rigor of the positivists that they criticized their own original program and admitted when it had failed. The somewhat more tolerant resulting position that gave up the strict verification criterion called itself **logical empiricism**. Verification was weakened to confirmation or partial support by the logical empiricists. Empirical criteria of meaning tended to be either too narrow, and exclude the more theoretical parts of science, or so broad as to succeed in allowing the more theoretical parts of science, but only to readmit metaphysics and theology to the realm of the meaningful. The original version of the verification principle excluded theoretical terms in physics, such as "electron," as meaningless, while the principle would allow "Either this object is red or God is lazy" to be verified by a red object and hence allow "God is lazy" to be meaningful.

Karl Popper (1902–94) was another Viennese philosopher of science who knew the logical positivists personally and debated with them, but whose views differ from theirs in some important ways. Popper claimed that falsifiability, the possibility of being falsified or refuted, not verifiability, is the criterion that separates science from non-science. This is Popper's **falsifiability criterion** of demarcation of science from non-science. Popper also claimed that the more falsifiable a theory is the more scientific it is. This leads to the view that scientific laws rather than statements of particular facts are the most scientific statements. (For the positivists statements of particular facts are the fully verifiable ones, hence the most scientific.) Particular facts can be verified, and, hence, achieve the highest probability, while laws cover an indefinite range of cases and can never be verified. In fact laws are always infinitely improbable according to Popper, because their range of application is infinite, but only a small part of their consequences is tested. The positivists' view of science fits with the view of science as primarily a collection of facts, organized and helpfully summarized by laws, while Popper's view of science makes science primarily a collection of laws. For Popper the role of particular facts lies solely in testing or attempting to falsify the laws. According to him science consists of bold conjectures or guesses and decisive refutations or negative tests. This is his **falsificationist method** of science. Popper accepts Hume's claim that there is no justification of induction, so Popper throws out induction as a "myth" (see box 1.1). Hypotheses are guessed at, conjectured, and do not arise logically from observations of individual cases. As long as a hypothesis survives testing it is scientifically retained. It doesn't matter whether it was preceded by observations or arose from a dream or a religious belief as long as it survives tests. The famous story of how the chemist F. A. Kekulé's (1829–96) dream of a snake swallowing its own tail led him to hypothesize the ring structure of the benzene molecule (Beveridge, 1957, 56, 76) is an example of how highly non-rational sources can nonetheless yield testable results.

Popper, unlike the positivists, did not equate non-scientific or non-testable with meaningless. For Popper metaphysics can be meaningful and can play a positive role in guiding the formation of scientific theories. Popper's views are counterintuitive at first, but their consequences fit well with the role of testing and criticism in science and the notion of the centrality of universal laws in science.

Popper's approach also has political implications. The critical approach (a generalization of the method of refutation) is central to free thought and democracy, the "open society" (Popper, 1945). The holding of positions as tentative avoids dogmatism. The welcoming of criticism encourages open-mindedness and free speech. Popper understood "reified dogmatism" as a closed system of thought that has mechanisms to discount or dismiss over possible objection or criticism. For Popper both religious fundamentalism and dogmatic Marxist totalitarianism are examples of such closed systems. However, Popper believes that schools of science themselves can develop strategies to shield themselves from all criticism, and thereby it too can become non-scientific in a logical sense, while mistakenly being held to be "science" by educational and funding institutions.

To take an extreme example, the psychologist of intelligence Sir Cyril Burt (1883–1971) was a leading **scientist in the institutional sense**. He edited the leading and most rigorous *Journal of Statistical Psychology*. He advised the London Council on educational tracking policy. He started Mensa, the society for people with high IQ, and even received a knighthood for his work on the inheritance of IQ. However, shortly after his death, most psychologists became convinced that the data he presented in his later years was fraudulent. He apparently invented non-existent research assistants. He composed letters and articles under pseudonyms incompetently criticizing his work, to give himself the opportunity to respond brilliantly to them (Hearnshaw, 1979). If this is true, he certainly failed in being a **scientist in the normative sense** of Popper; that is, someone who was intellectually honest, open to criticism, and willing to reject his theories.

The interesting aspects of Popper's contrary to common sense approach to science were only widely recognized after logical positivism ran into conceptual difficulties. Popper's views have exciting implications for both social criticism and evaluation of programs in science, but the downside of Popper's views for the philosophy of technology is the sharp wedge he drives between science and technology. Science involves daring and improbable conjectures and their refutation, but technology demands reliable and work-able devices. The collapse of a bridge has human costs different from those of intellectual rejection of a theory in particle physics. Popper students and followers, such as Joseph Agassi (1985) and Mario Bunge (1967, chapter 11, 1979), have made important contributions to the philosophy of tech-nology, but Popper's own theory of science, interesting as it is, is separate from pragmatic considerations of technology. Nevertheless, the Popperian approach to science opens the way to investigation of the way that philo-sophical worldviews or metaphysical theories may influence the formation of scientific theories. This in turn shows how cultural views can be at least as important as observational data as a source of scientific theory, and through applications can in turn affect technology.

One of the major debates in the philosophy of science is that between **realism** and **anti-realism**, particularly, as Popper (1962, chapter 3) formu-lated it, between essentialism and **instrumentalism** with respect to theoret-ical terms in science. Some parts of science are particularly close to observation and experiment. Other parts of scientific theory are only indirectly con-nected with observation and experiment through long chains of logical deduction. The term "electron" in physics is an example. Scientific realists claim that the theoretical terms in science represent or refer to objectively real entities, even if we cannot observe them. Anti-realists claim that the theoretical terms are not to be taken to literally refer to objects or entities.

Instrumentalists treat theories merely as instruments for prediction. The theories do not describe real, unobserved, structures, but are more or less useful for prediction of things we can directly observe.

The metaphors used by realists and instrumentalists, respectively, are at the basis of the theoretical and technological approaches that were histor-ical components of the Renaissance birth of modern science. Realists often describe scientific theories as "pictures" of the world. Instrumentalists describe theories as "tools" for prediction. The birth of early modern science may have been the fusion of literate scholars, knowledgeable of the Greek clas-sics and philosophical "world pictures," but ignorant of practical crafts, with the illiterate but technologically skilled artisans with their tools. Economic hard times in the Renaissance threw together impoverished wandering scholars with wandering artisans, yielding the "marriage of metaphysics and technology" (Agassi, 1981) that is science. The claim that the breakdown of class barriers led to communication between humanists and artisans (Zilsel, 2000) is called "the Zilsel thesis." While Zilsel (1891–1944) located this process in the 1600s, it is more plausible to trace it to the Renaissance in the 1400s

(Rossi, 1970). The "Renaissance Men," such as Leon Battista Alberti (1404–72), theorist of geometrical perspective, architect, and social philosopher of the family, and Leonardo da Vinci (1452–1519), the artist, philosopher, scientist, and engineer, were artists who combined technical virtuosity in mechanics and architecture with philosophy and scientific theory. The "tools" in the form of painter's brush or sculptor's chisel were a means to the "picture" in the form of Renaissance painting or sculpture. These components of the birth of modern science and technology show up today in the preferred metaphors of instrumentalism and realism as philosophies of the relation of science to reality.

The interesting aspects of Popper's non-commonsensical approach to science were only widely recognized after logical positivism ran into conceptual difficulties. During the 1950s and 1960s there were many criticisms of logical empiricism. The logical empiricists were rigorous and honest enough to qualify and limit many of their own claims. The history of logical empiricism is one of successive retreat from the original, simple, and provocative theses of the Vienna Circle. This whittling down of logical empiricism greatly increased interest in and allegiance to Popper's alternative approach.

However, the most well known and influential alternative was Thomas Kuhn's *The Structure of Scientific Revolutions* (1962). Kuhn (1922–96) approached science from the standpoint of history. Kuhn, with a doctorate in physics, taught an undergraduate course in science for humanists, reading original texts. Kuhn was puzzled by Aristotle's *Physics*, which seemed totally nonsense to someone trained in modern physics. One afternoon, while he was gazing from his dorm window out on the trees of Harvard Yard, the scales fell from his eyes, and he realized that Aristotle's claims made perfect sense within a framework totally different from the modern one.

The logical positivists had treated scientific theories as static structures. They made their own formal reconstructions of scientific theories, rather than describing the theories as their creators and contemporaries viewed them. Kuhn claimed to present scientific theories in terms of the frameworks in which they were originally understood, not as contemporary textbooks or logical empiricist formal reconstructions presented them. Kuhn centered his account of science on the notion of a **paradigm.** The paradigm is not solely an explicit formal structure. Paradigms are not only explicit theories but ways of viewing the world. Kuhn's paradigms include not only (a) theories, but also (b) tacit skills of laboratory practice that are not recorded and are taught by imitation of an expert practitioner. Further, paradigms encompass

(c) ideals of what a good scientific theory should be, and (d) a metaphysics of what basic entities exist. Kuhn also tied the paradigm to the structure of the scientific community. The paradigm binds the researchers in a scientific specialty, channeling their experimental and theoretical practice in certain directions and defining good scientific theory and practice. Later Kuhn distinguished between paradigms as **exemplars**, models for good scientific theory and practice, such as the works of Galileo, Newton, or Einstein, and the paradigm as **disciplinary matrix**, or belief system shared by members of the scientific community.

Kuhn's view of the development of scientific paradigms differs from the positivists' and Popper's account of theories. Kuhn denies that either induction or Popper's falsification describes the rise and fall of paradigms. Generally a new paradigm arises without a strong inductive base. Particular refutations can be sidestepped by modifying one or more hypotheses in the theory refuted. The scope of the original theory can be limited or auxiliary assumptions can be added. Thus "refutations" are not decisive or fatal. A slightly modified version of the "refuted" theory can survive under the paradigm. The logic of this situation is called the **Duhem thesis** or **Duhemian argument** (see box 1.2).

Paradigms collapse because of an accumulation of what Kuhn calls **anomalies**. Anomalies are not strict counter-instances or refutations. They are phenomena that seem not to fit with the categories of the paradigm or that are simply left as exceptions. A paradigm is rejected only after a new paradigm has arisen and there is a shift of allegiance of scientists. (At one point Kuhn quotes the physicist Max Planck as saying that it is a matter of the older generation dying off; Kuhn, 1962, p. 151.)

Kuhn's approach opened the way for widespread appreciation of the role of philosophical worldviews and social ideologies in the creation and acceptance of scientific theories. Kuhn himself did not emphasize either the philosophical frameworks of theories as such or the external, social influences on the acceptance of new paradigms, though he hinted at both in passing. However, after Kuhn, numerous philosophers, historians, and sociologists of science took up the issue of how philosophical views, religious beliefs, and social ideologies have played a role in the birth and spread of scientific theories. This in turn strengthens the case for cultural influences on technology. If the scientific paradigms at the basis of various technologies have religious or political components, then religion and politics can influence technology, not just in social acceptance, but also in terms of the structure of the very theories used in the technology. This approach, taking its

Box 1.2

The Duhem thesis

The Duhem thesis is the logic behind much of Kuhn's claim that paradigms in normal science do not get rejected because they are refuted. The physicist, philosopher, and historian of science Pierre Duhem (1861–1916) presented arguments against crucial refutation of theories in the early twentieth century. He wrote decades before Popper but his arguments are the most challenging objection to Popper's theory of falsification. Duhem noted that if a theory consists of several hypotheses or assumptions, the refutation of the theory as a whole does not tell us which hypothesis is a fault, only that the theory as a whole made an incorrect prediction. Duhem also argued and gave examples, in which one of the lesser hypotheses or auxiliary hypotheses was changed in the light of a supposed refutation of the theory, yielding a modified theory that correctly described the situation that refuted the original version. For instance, Boyle's law of gases appeared to be refuted by the behavior of iodine. However, chemists and physicists modified Boyle's law to claim that it applied only to ideal gases, and then claimed that iodine was not an ideal gas. (In an ideal gas all molecules are the same. Iodine gas is a mixture of molecules with different numbers of iodine atoms.)

The American philosopher W. V. O. Quine (1908–2000) generalized Duhem's claim, noting that if one allows drastic enough modifications and redefinitions *any* theory can be saved from any counter-evidence (Quine, 1951). Quine allows such extreme strategies as changing the formal logic of the theory and "pleading hallucination" (the extreme Quinean strategy could also, more reasonably, involve shifts in the meaning of terms in the theory to avoid refutation). This is known as the *Duhem–Quine thesis* and has influenced constructivist sociologists of scientific knowledge and science and technology studies postmodernists. The latter claim that, because no evidence can decisively refute any theory, the reason why theories are rejected involves non-evidential issues, not part of the logic of science, such as political, social, or religious interests and worldviews, and these views in turn can influence the technology based on the theory.

Philosophers have disagreed about the legitimacy of various stronger and weaker versions of the Duhemian argument (Harding, 1976). Analytical philosophers of science generally defend the weaker, original, Duhem version of the thesis. Science and technology studies people generally opt for the strong, Quine, version, as it appears to make evidence almost irrelevant to science (see box 6.3 and the discussion of the sociology of scientific knowledge in chapter 12).

inspiration from Kuhn, has been used to counter technological determinism (see chapter 6).

Two topics raised by the post-positivist philosophy of science are the **theory-laden nature of observation** and the **underdetermination** of theories by evidence. Kuhn, as well as several other philosophers of science of the late 1950s and 1960s, such as Norwood Russell Hanson (1958), emphasized how sensory observations depend on contexts of theory and interpretation. They appealed to psychological studies of perception and visual illusion. They followed the claim that beliefs and expectations influence perception. Michael Polanyi (1958) emphasized how skills of interpretation are developed though apprenticeship and practice. The interpretation of medical X-ray pictures or the identification of cell organelles through a microscope is not obvious. It involves training. (James Thurber recounts how, after a long bout with a school microscope, he realized he was studying not a microscopic creature, but the reflection of his own eyelash.)

Another form of theory dependence of observation includes the role of theories of our measuring and observing instruments in our construal of the readings and observations our instruments produce. Theory also plays a role in the selection of what to observe and in the language in which observations are described and interpreted. Even when perceptual observation has been replaced by machine observation these latter influences of theory on the nature and structure of observation reports remain.

Closely related to the problem of induction (box 1.1) and to the Duhem thesis (box 1.2) is the **underdetermination thesis**. Many different theories, such as both the new theory and the appropriately modified version of the old, supposedly refuted theory, can explain the same set of data. The same set of data points can be predicted or accounted for by many different equations. Many different continuous curves can be drawn through any set of points, and thus the many different equations of those curves can be said to describe those points. Thus the logic of confirmation or inductive support does not lead to one unique theory. Considerations other than empirical evidence are used in making the choice of theory. It is true that most of the mathematically possible theories that account for a given set of data are excessively complicated and most could be eliminated as unreasonable. Nevertheless, where more than one reasonably manageable theory accounts for the data, there is a turn to considerations of simplicity or elegance. But what is considered simple depends on ideals of what a good theory is and aesthetic considerations of the scientist or scientific community. Consistency with other theories held also counts as a non-empirical constraint on theory choice.

Many sociologists of scientific knowledge and science and technology studies postmodernists appeal to the underdetermination thesis as well as the Quine–Duhem thesis.

Sociology of Scientific Knowledge

The work of Kuhn (and a number of other philosophers of science of the period) opened thinking about science to a number of issues and considerations that the positivist approach to science as a body of statements had overlooked or not considered worthy of investigation. Kuhn's paradigmatic approach opened science to the kinds of examination that humanists had traditionally used in examining art and culture. It also opened up the social examination, not just of the institutions and networks of science, but also of the content of scientific theories (sociology of scientific knowledge or SSK) (see chapter 12). A number of sociologists of science, particularly in the United Kingdom, took up this investigation.

Earlier sociology of knowledge, initiated by Karl Mannheim (1893–1947) in 1936, had investigated political and religious beliefs but exempted the claims of science from sociological explanation (Mannheim, 1936, p. 79). A majority of sociologists consider themselves to be scientists and to share at least a diluted version of Comte's (1798–1857) positivistic ideal of objective, lawful sociological knowledge (see the discussion of Comte in chapter 3). Most sociological studies of science prior to the 1970s concerned networks of journal citations or patterns of funding and professionalization, assuming the content of science to be beyond social explanation. Robert K. Merton's (1910–2003) influential sociology of science concentrated on the norms of science. These are the values professed by the scientific community. These norms include: (a) universalism; (b) disinterestedness (lack of interest-based bias in investigation); (c) "communism" (the free sharing of data and results); and (d) organized skepticism (the tendency to doubt and question results and theories). These values resemble Karl Popper's norms of science, although the later Popper emphasizes that these are ideals of science, not a description of actual scientists' behavior. (In contrast to Popper, Kuhn claims to be giving an account of the actual behavior of scientists.) Philosophically trained sociologist of science Steve Fuller (1997, p. 63) has noted that Merton takes the professed ideals of scientists at face value as accounts of actual scientific behavior, while sociologists of politics and religion often doubt or even debunk explicitly professed ideals and contrast them with the actual

17

doings of politicians and religious people. It is significant that Merton first discussed the norms of science in an article concerning totalitarian restraints on science in Nazi Germany (Merton, 1938). His use of the term "communism" for sharing of data shows a residue of his earlier leftist political views. Later, Merton contrasted these norms of science primarily with those of the USSR.

SSK claims that the statements, laws, and experiments of science are themselves legitimate objects of scientific investigation. Earlier philosophers and sociologists of science thought (and many still think) that scientific errors could be explained by social or psychological causes, but scientific truths could not. David Bloor (1976), who initiated what he called the "Strong Programme," proposed (a) a Principle of Symmetry, that the same sort of causal explanations ought to be given of both truth and error in science and of both rational and irrational behavior; (b) a Causality Principle, that all explanations of scientific knowledge should be causal; (c) the Impartiality Principle, that SSK should be impartial with respect to truth and falsity, rationality and irrationality; and (d) the Reflexivity Principle, that these principles should apply to sociology itself.

Other sociologists of scientific knowledge such as Harry Collins (1985) bracketed or set aside the truth or falsehood of scientific statements, studying gravitational waves and parapsychology with the same methods and approaches. Many in SSK followed the social construction of scientific knowledge approach (see chapter 12). The social construction approach can mean a number of things. The weakest version of the thesis is that social interaction of humans is the basis for the formation of scientific theories and experiments. This claim is reasonable. Science differs from introspective knowledge in that it is supposed to be public and replicable. Science is a social enterprise. Another reasonable claim is that technological, instrumental apparatus is physically constructed in a literal sense. An issue arises, however, as to whether the construction of concepts and the construction of apparatus ought to be considered with a unitary conception of "construction" or whether two different sorts of activities are being illicitly run together.

A stronger thesis of the social construction position is that the objects of science or scientific truths are socially constructed. If the latter claim means that what we take to be scientific truths or what we believe to be scientific truths are socially constructed then it collapses into the first position. Many of the defenders of social construction would claim that there is no difference between a truth and what we take to be a truth. This is a version of the consensus theory of truth, that what is true is what the community believes. Opponents of the extreme social construction view claim that the objects that we think exist may not be the same as the objects that genuinely exist and that what our community holds to be true may not be true.

Among the contributions of the social constructivist and related approaches are case studies showing how consensus is formed in scientific communities. Political negotiation, appeals to authority of eminent scientists, recruitment of allies, and rhetorical persuasion of the undecided all play a role. Extra-scientific factors often play a role. Pasteur's experiments in the mid-nineteenth century rejecting the spontaneous generation of life from non-living material were welcome to the Catholic Church in their defense of the necessity of divine creation. Although Pasteur himself did not really reject an origin of life in the distant past due to natural causes, he was happy to play to the conservative atmosphere in France in his day (Farley and Geison, 1974; Geison, 1995). Often the consensus is formed without some major objections being accounted for. Some conflicting experiments and studies are rejected because of the reputation or lack of prestige of the experimenters. Data that don't fit the expected result are ignored. Once the conclusion is reached it seems in retrospect to be inevitable. It is difficult to remember or imagine the state of uncertainty and disagreement that preceded it.

Social Epistemology

In Anglo-American analytical philosophy a field of **social epistemology** (social theory of knowledge) developed in the last two decades of the twentieth century (Fuller, 1988; Kitcher, 1993). Social epistemology, like traditional epistemology, but unlike sociology of scientific knowledge, is normative. That is, it is concerned with evaluation of the knowledge claims made by scientists. On the one hand, social epistemology differs from traditional philosophical epistemology in treating knowledge, especially scientific knowledge, as a social rather than an individual achievement. On the other hand, social epistemology contrasts with Kuhn's and other "historicist" (history-oriented rather than logic-oriented) post-positivistic philosophy of science in claiming to take a descriptive approach. It also contrasts with the evaluatively neutral stance taken by most social constructivist sociology of science, such as in the bracketing of truth-claims by Collins noted above. Some "historicist" philosophers of science, such as Feyerabend, do take evaluative stances, in judging and rejecting certain scientific theories, and social constructivists, while professing a neutral stance, often implicitly debunk the traditional "naive" truth claims made by scientists. That constructivists are not as totally normatively neutral as they profess to be is suggested by the fact that though they treat occultism or parapsychology neutrally, none so far has treated racist science or Nazi science in an evaluatively neutral manner.

Feminist, Ecological, and Multicultural Science and Technology Studies

Accepting Kuhn's views, if cultural considerations of worldview and attitude toward nature were significant for the nature of scientific theories, then theorists of science with critical approaches to dominant social attitudes could criticize the theories and methods of various sciences and technologies themselves. Feminists and ecological critics of contemporary society have been prominent in taking this approach (see chapters 9 and 11). Similarly, anthropology and some cultural studies approaches to science and technology have criticized the assumption of universality of Western science and technology. These responses note earlier, but often fruitful, approaches to science and technology in Egyptian, Babylonian, Chinese, Indian, and Muslim civilizations of ancient and medieval times. These cultures contributed a great deal of technology and science to the West, but often based their investigations on worldviews and metaphysics very different from that of modern Western science. From this multicultural critics raise questions about the supposed "universality" of Western science (Harding, 1998).

Likewise, the "local" knowledge in non-literate societies has often contained considerable knowledge of medicinal and other values of local plants, agricultural techniques, survival skills in harsh climates, and navigational skills. Contemporary ethno-botanists investigate indigenous cures and the chemistry of plants used by local healers. Western Arctic explorers borrowed the design of their clothing and many survival techniques from the Inuits and other inhabitants of the Arctic, usually without crediting them. Apparently local religiously based seasonal cycles of planting, such as those in Bali, have sometimes proven more agriculturally effective than the recommendations of Western "experts." Social constructivist and postmodernist defenders of ethno-science sometimes claim that it is simply an alternative knowledge to Western science, which is itself a "local knowledge," whose locality is the laboratory. (See chapter 10.)

The "Science Wars"

With the post-Kuhnian development of politically critical studies of science by feminists, activists in indigenous cultures, ecologists and others, as well as the development of sociological studies of scientific knowledge and literary studies of scientific texts a backlash has arisen. A number of disparate groups are involved. There are scientists and technologists who believe the objectivity of their field is being wrongly denied by social, political and literary studies of science. There are also political conservatives, opponents of feminism and ethnic movements, and opponents of the ecology movement. There are also traditional literary and humanist opponents of postmodernist movements in the humanities (see box 6.3). These groups have formed an unstable alliance to attack the new science studies in the so-called **Science Wars** (Ross, 1996; Dusek, 1998). A great many articles, both scholarly and polemical, were written for and against the new science studies (Koertge, 1997; Ashman and Baringer, 2001). The most famous, or notorious, incident in the science wars was the **Sokal hoax**. Alan Sokal, a physicist, wrote an article titled "Transgressing the boundaries: toward a transformative hermeneutics of quantum gravity" (1995), which contained within its implicit satire the most ridiculous and exaggerated claims to be found in science studies and the political criticisms of science. He was able to publish it in a cultural studies journal, and then revealed his hoax. In the wake of this revelation both sides in the dispute issued hundreds of news stories, editorials, and articles both pro and con, ranging from conservative political commentators Rush Limbaugh and George Will to angry letters to editors of various eminent scientists and humanists (Editors of *Lingua Franca*, 2000). At least two science studies people were denied prestigious positions, and an editor nudged into earlier retirement, because of letter writing campaigns by the scientist science warriors. The fires of the science wars in the form that gained mass media attention died down by the beginning of the new century, but they continue to smolder in less public and explicit forms.

Instrumental Realism

A later development in the philosophy of science with the greatest relevance to philosophy of technology was what Don Ihde (1991, p. 150 n1) called the **instrumental realist** approach to science. The positivists, the followers and descendents of Kuhn in the new philosophy of science, and even the sociology of scientific knowledge concentrated on science as primarily a *theoretical* enterprise. Empirical testing was definitive of positivism's definition of scientific knowledge, but the model of testing was generally direct sense observation.

Philosophers and historians such as Ian Hacking (1983), Robert Ackermann (1985), and Peter Galison (1987) emphasize the mediating role of observational instruments and the manipulative nature of scientific knowledge. The American pragmatists, such as John Dewey, early emphasized physical manipulation of nature as central to knowledge. However, later academic pragmatism became strongly influenced by the positivists. Emphasis on practice and manipulation declined in later "pragmatic" accounts of science.

Kuhn had included tacit laboratory skills in his account of science, but later discussions of Kuhn in philosophy of science debates focused primarily on the conceptual aspects of the paradigm. Jerome Ravetz developed Kuhn's emphasis on the craft nature of science and Michael Polanyi's notion of tacit skills into a thoroughgoing treatment of science as craft, but it was not influential, perhaps because it was neither fully within philosophy of science nor within the developing sociology of scientific knowledge (Ravetz, 1971). The instrumental realists of the 1980s developed a strong focus on the embodied, active aspects of science that became a significant movement within the philosophy of science. For the instrumental realists scientific instrumentation is central to science. The active, manipulative aspect of instrumental observation takes priority over passive observation and contemplation. Today most scientific observations are far from the naked eye observations of older astronomy and natural history. The "Baconian" ideal of induction from pure, unbiased perceptions is modified by the technology-laden nature of contemporary scientific observation. French physicists call the tendency to look where our instruments allow us to look "the logic of the lamp post," after the old joke about the drunk who looked for his keys under the lamp post because there was better light, even though he had dropped his keys down the block where it was pitch black.

Since contemporary science is so involved with and dependent upon sophisticated technological instrumentation, insofar as scientific discovery is based on observation, technology is prior to science as well as driving science. This is the opposite of the account of technology as "applied science," in which science is prior to and drives technology. In the "technoscience" view of Bruno Latour (1947–) and others, technology and science are today inextricably interwoven. The notion that modern science is dependent on technology has some similarity to Heidegger's later views (see box 5.1). For Heidegger, technology is the fundamental fact or force in the modern condition, and technology is philosophically prior to science.

Instrumental realism shifts the dividing line between theory and observation in such a way as to make the realm of pure theory minimal. Where the ability to manipulate the situation is a criterion of the reality of the entity manipulated, much that was formerly treated as "purely theoretical" in the philosophy of science becomes real. Hacking's famous example is that on hearing that elementary particles (often treated as theoretical entities by philosophers) can be sprayed, he concluded if they can be sprayed they are real (Hacking, 1983). The realist/instrumentalist dispute mentioned earlier in this chapter treated theoretical entities as objects of contemplation, not of manipulation. By rejecting this contemplative stance instrumental realists not only make clearer the close connection of modern science with technology (thereby implicitly justifying the running together of the two as "technoscience" by postmodern science studies), but also eliminate the break between ordinary experience and the objects of science.

Approaching and integrating the work of the instrumental realists from a phenomenological point of view (see chapter 5), Don Ihde has drawn the implication of this approach that even the most esoteric scientific research involving advanced, abstract theories is highly perceptual, given that testing via instrumentation is an extension of perception. Ihde also points out that instruments as an extension of (or, better, literally incorporated into) bodily perception incorporate human embodiment into even the most arcane, advanced science.

One irony of the development of the instrumental realist approach from post-positivist philosophy of science is that in its application to the history of science the emphasis on experiment has reintroduced the characterization of scientific method as inductive. Yet it is precisely the problem of induction and responses to it such as Popper's that led to the emphasis on the theory-driven account of science and on the theory-laden character of observation.

Historians and sociologists of science who use the instrumental realist approach may not be bothered by these problems (though at least one sociologist of scientific knowledge, Collins, uses the problem of induction to undermine empirical accounts of scientific change). Instrumental realism brings us full circle to the original inductivism, if not that of the simplest variety. Philosophers, however, may need to re-examine their approaches to the logical problem of induction in relation to the defense of the instrumental realist approach.

Conclusion

Inductivism supported the view that science grows directly out of perceptual observations unbiased by theory. The logical positivist or logical empiricist philosophy of science has often been used to reinforce the notion of science as neutral and technology as applied science. Popper's falsificationism or critical approach, belatedly appreciated, allowed for the role of theory as prior to observation and the role of philosophical theories as background frameworks for scientific theories. Kuhn and post-positivist, historicist philosophy of science opened the door to considering the role of philosophical, religious, and political influences on the creation and acceptance of scientific theories. Feminist, ecological, and multiculturalist critics use Kuhn's notion of a paradigm to expose what they claim to be pervasive bias in the methods and results of mainstream Western science and technology. Sociologists of scientific knowledge emphasize that the logic of evidence and refutation of theories does not determine the course of theory change. Instead, the prestige of established scientists, recruitment of allies, and negotiation between competing teams leads to a closure of scientific disputes that is later attributed to the facts of nature. Instrumental realists emphasize that, in contemporary science, observation itself is mediated through the technology of scientific instrumentation. Rather than technology being applied science, technology is prior to scientific observation.

Study questions

1 Do you think that inductivism is adequate as a theory of scientific method? If it is not, why do so many working scientists hold to it?
2 Are scientific theories decisively refuted by counter-evidence or do they "roll with the punches," so to speak, being readjusted to fit what was

counter-evidence for the old version? Give an example not in the chapter if possible.

3   Are scientific theories direct outgrowths of observation or are they influenced as well by assumptions and worldviews of their creators?

4   Do you think that the scientists involved in the "science wars" who dismiss and ridicule sociological and literary accounts of the success of scientific theories are justified in dismissing the social, political, and rhetorical aspects of science as irrelevant to scientific truth and validity?

5   Does the instrumental realist approach do away with the problems raised by earlier accounts of science, such as the inductive, Popperian falsificationist, and Kuhnian approaches?

2

What Is Technology? Defining or Characterizing Technology

Why Bother with Definitions?

Many students, in my experience, especially in the natural sciences, are impatient with disputes about definitions. They are often called "merely semantic" and may seem hairsplitting. Indeed, they are semantic, in that they deal with meaning, but they are hardly trivial. Many apparently substantive disagreements really stem from the disputants having two different definitions of what is being discussed, say religion, but not being aware of it. Often people think that definitions are purely arbitrary; it means that effort need not be wasted on choosing among opposing or alternative definitions. This is itself based on one view of definition, but it is not the only one. We shall learn something about philosophy by seeing the different sorts of definitions that people have used and their connection to differing philosophical views.

Looking at the alternative definitions of technology shows something about the alternative kinds of definition and also about the characterization of technology. Even if one doesn't find a final definition on which everyone can agree, an investigation of the definition of technology shows us the range of things that can count as technology and some of the borderline cases where people differ on whether something should be counted as technology or not. Even an unsuccessful search for a best definition helps us to explore the layout of the area we are investigating.

Kinds of Definitions

Let us look at a few different sorts of definitions. At one extreme is the ancient notion of a **real definition**. The ancient Greek philosophers Socrates

WHAT IS TECHNOLOGY?

(470–399 BCE), his student Plato (428–347 BCE), and his student, in turn, Aristotle (384–322 BCE) held to this notion of definition. This view assumes that there is a real structure to the world that corresponds to our words and that a correct definition will match the real nature of things. Socrates went about questioning people about the definition of notions such as justice, courage, or piety and showed the people he questioned how their definitions failed to fit with their notions. Socrates appeared to assume and Plato argued that there is a real nature or structure of justice, of courage, and of piety and that the real definition will fit this. Aristotle claimed that objects have essences and that real definitions will match these. Real definitions of the sort that Plato and Aristotle sought are supposed to "cut nature at the joints"; that is, correspond to the "natural kinds" of things. Some contemporary thinkers view scientific definitions, such as definitions of the chemical elements in terms of atomic weight and number, as true definitions in this sense. Some recent writers in technology studies claim that leading writers on technology in the twentieth century, such as Martin Heidegger (1889–1976) and Jacques Ellul (1912–94), are mistakenly searching for an "essence" of technology. Heidegger, in fact, rejected the traditional account of forms and categories of Plato and Aristotle. Nevertheless, it is true that Heidegger and Ellul do present what they claim is a single, real, core notion of technology.

A different, nearly opposite, view of definitions is that of **stipulative definitions**. This conception is closer to the view of definitions held by many people today. It is claimed that definitions are arbitrary choices or stipulations. Definitions are about words and not things. Opponents of the notion of real definitions deny that there are natural classes or real natures of things to be captured by definitions.

Definitions, on the nominalist view, arbitrarily carve up the world of individuals into classes of things. One can define anything as anything one wishes. Lewis Carroll (1832–1898), who was a logician as well as a writer of children's books, had Humpty Dumpty hold this view of definitions. Humpty claimed it was a matter of who was master, he or the words. But as some of Humpty's definitions showed, we cannot sensibly define things in absolutely any way we want. We cannot define religion as a coffee pot and expect to make progress investigating the features of religion. In purely formal systems of abstract math or logic, stipulative definitions make more sense than in common sense or everyday discussions. In an abstract system of math one can lay down a definition and carry it through the system by exact rules of inference. One can use stipulative definitions for the purposes of argument or for a very limited investigation of everyday concepts, but one problem

Box 2.1

Nominalism in British philosophy

Nominalists in the late Middle Ages, such as William of Ockham (1285–1347), denied the reality of essences or universals and claimed that only individuals are real. British philosophers of the seventeenth and eighteenth centuries, in the early days of experimental science, held that there were no real definitions. Thomas Hobbes (1588–1679) claimed that definitions are stipulative, even though he thought he could base science on them. In the seventeenth century, Hobbes successfully described the definitional or postulational side of science, but he failed to explain how he tied his definitional and deductive notion of science to observation. For Hobbes, definitions are introduced at the start of an investigation, they are not, as they were for Aristotle, the final result of investigation. John Locke (1632–1704) claimed that we cannot know the real essences of substances. Definitions do not describe essential properties of things or even whether the things defined exist. We can know only nominal essences of substances. In the early eighteenth century David Hume (1711–76) totally denied the existence of real essences, and his position was highly influential upon later empiricism.

with using stipulative definitions in everyday reasoning is that the ordinary meaning of commonly used words sneaks back into the discussion without the author noticing. She slides from her stipulative definition to the ordinary meaning unaware. Writers on technology are, of course, free to define it any way they wish, but they then need to be careful they do not slip back into using other definitions or understandings of technology common in the culture without realizing they have strayed from their original definition.

This leads us to another sort of definition different from both the above, the **reportative definition**. This sort of definition is a report of how people ordinarily use words. It doesn't claim to find the true structure of reality, but it also doesn't simply make up an arbitrary definition by fiat. Dictionary definitions are close to reportative definitions. However, a purely reportative definition would simply describe how people use the word, without legislating "proper" usage. Dictionary definitions contain some normative content. A pure reportative definition could be quite complicated, describing how people in different regions or of different social status use the word. Reportative definitions often have fuzzy boundaries or vagueness of application. Ordinary language is frequently imprecise as to exactly what objects count as falling under the definition. The problem with using reportative definitions of technology is that there are so many different uses of the term around. For instance, some educators associate the word "technology" solely with computers in the classroom, while the school building itself, as well as such older aids to teaching as the blackboard, are part of technology in the broadest sense.

A kind of definition used in philosophy and in other academic areas is a **précising definition**. This sort of definition retains the core ordinary meaning of the word. It is not stipulative or arbitrary. However, unlike a reportative definition it does not simply describe how people actually use the word. It attempts to sharpen up the boundaries of application of the word by describing the range of application and cut-off points. (How big is "big" for a certain kind of thing? How few hairs can one have and still be counted as bald?) Any philosophical attempt at a general definition of technology will be a précising definition.

British empiricist philosophers rejected the existence of essences as real natures of things, but the notion of definition by single defining characteristic continued in general use. In the second half of the twentieth century a number of philosophers concluded that kinds of entities couldn't be characterized by an essence. One view (of Ludwig Wittgenstein, 1889–1951) is that objects classified under a single name do not share any one single characteristic but share a "family resemblance." One can often recognize similarities between members of the same human family but cannot find any single feature that they all share. Any pair of things in the class shares some characteristics but no one characteristic is shared by all of them. Wittgenstein gave the example of the notion of "game." The usual characteristics used to define a game are not shared by every game. Not all games have competing players. Not all games have hard and fast rules, involve equipment or game pieces, and so forth. A game is best defined according to the family resemblance approach by giving paradigmatic examples and suggesting that similar things should also be included. Some contemporary philosophers present views that would make technology an example of a family resemblance concept. Current philosophers of technology such as Don Ihde, Donna Haraway, Andrew Feenberg, and others have ceased searching for an "essence of technology" of the sort propounded by early or mid-twentieth-century thinkers such as Martin Heidegger (see chapter 5) or Jacques Ellul (see below in this chapter, and chapter 7). It is suggested that the things that are included under technology are too varied and diverse to share a single essence.

As mentioned above, the major theorists of technology of the first two-thirds of the twentieth century believed that a universal, essential definition of technology could be given. A number of recent theorists, such as Don Ihde, Andrew Feenberg, and others, believe, in contrast, that there is not an essence or single defining characteristic of technology, and that searching for an essential definition is unproductive.

Guidelines for Definitions

Some general guidelines for definition are the following:

1   A definition should not be too broad or narrow. (That is, the definition should not include things we would not designate by the word we are defining, and the definition should not be so restricted as to exclude things that should fall under the term defined.)
2   A definition should not be circular. (For instance, we shouldn't define "technology" as "anything technological" and then define "technological" as "anything pertaining to technology.")
3   A definition should not use figurative language or metaphors.
4   A definition should not be solely negative but should be in positive terms. (A purely negative definition in most cases would not sufficiently limit the range of application of the term. A definition by contrast has to assume that the hearer knows the contrasting or opposite term.)

Box 2.2

Philosophical exceptions to the standard rules for definition

The rough guidelines for definitions will have exceptions if one holds certain non-commonsensical philosophical views. For example, some mystics believe God can be characterized only negatively, and hold to so-called "negative theology." Although a simply circular definition is completely unhelpful, it has been pointed out that if one follows out the definitions in a dictionary, looking up the words in the definition, one eventually goes in a circle, although it is a big circle. Some philosophers, such as Hegel, have suggested that the point is not to avoid circularity but to make the circle big enough to encompass everything!

An example of defining technology in a too narrow manner is the common contemporary tendency to mean by "technology" solely computers and cell phones, leaving out all of machine technology, let alone other technology. A case of defining technology in a manner that may be too broad is B. F. Skinner's inclusion of all human activity in technology. Skinner understands human activity as being conditioned and self-conditioning. For Skinner conditioning is considered to be behavioral technology. A related move is the general inclusion of "psychological technology" as part of the motivational apparatus of technological activities, such as chanting in hunter-gatherer societies, or various political beliefs in industrial societies (propagated by propaganda, understood as a kind of technology by Ellul), thereby erasing the distinction between technology and culture by including *all* of culture within technology (see below on Jarvie).

Definitions of Technology

Three definitions or characterizations of technology are: (a) technology as hardware; (b) technology as rules; and (c) technology as system.

Technology as hardware

Probably the most obvious definition of technology is as tools and machines. Generally the imagery used to illustrate a brochure or flier on technology is that of things such as rockets, power plants, computers, and factories. The understanding of technology as tools or machines is concrete and easily graspable. It lies behind much discussion of technology even when not made explicit. (Lewis Mumford (1895–1990) made a distinction between tools and machines in which the user directly manipulates tools, while machines are more independent of the skill of the user.)

One problem for the definition of technology as tools or machines is cases where technology is claimed not to use either tools or machines. One such non-hardware technology is the behavioral technology of the psychologist B. F. Skinner (1904–90). If one considers verbal or interpersonal manipulation or direction of the behavior of another as technology then it appears we have technology without tools. Mumford claims that the earliest "machine" in human history was the organization of large numbers of people for manual labor in moving earth for dams or irrigation projects in the earliest civilizations, such as Egypt, ancient Sumer in Iraq, or ancient China. Mumford calls this mass organized labor "the megamachine" (Mumford, 1966). Jacques Ellul considers patterns of rule-following behavior or "technique" to be the essence of technology. Thus, propaganda and sex manuals will be technology involving rules, and can, but need not always, involve use of tools or hardware.

Technology as rules

Ellul's "technique" mentioned above is a prime example of another definition of technology. This treats technology as rules rather than tools. "Software" versus "hardware" would be another way to characterize the difference in emphasis. Technology involves patterns of means–end relationships. The psychological technology of Skinner, the tool-less megamachine of Mumford, or the "techniques" of Ellul are not problems for this approach to technology. The sociologist Max Weber (1864–1920), with his emphasis on "rationalization," resembles Ellul on this, characterizing the rise of the West in terms of rule-governed systems, whether in science, law, or bureaucracy. Physical tools or machinery are not what is central; instead it is the means–end patterns systematically developed.

Technology as system

It is not clear that hardware outside of human context of use and understanding really functions as technology. Here are some examples:

1   An airplane (perhaps crashed or abandoned) sitting deserted in the rain forest will not function as technology. It might be treated as a religious object by members of a "cargo cult" in the Pacific. The cargo cults arose when US planes during the Second World War dropped huge amounts of goods on Pacific islands and cults awaited the return of the big "birds."
2   The Shah of Iran during the 1960s attempted to forcibly modernize the country. He used the oil wealth to import high technology such as jet planes and computers, but lacked sufficient numbers of operators and service personnel. It has been claimed that airplanes and mainframe computers sat outside, accumulating sand and dust or rusting, as housing for storage and the operating and repair staffs for them were not made available. The machinery did not *function* as technology.
3   Technological hardware not functioning as technology is not solely the province of indigenous societies or developing nations, but can also be

present in a milieu of high tech, urban sophisticates. Non-Western technology was displayed in an exhibit of "Primitive [*sic*] and Modern Art" at the Museum of Modern Art as purely aesthetic or artistic phenomena. Indigenous implements and twentieth-century Western abstract art objects were exhibited side by side to emphasize similarity of shape and design. The labels of the primitive implements often did not explain their use, only their place and date. (The use of these devices for cooking, navigation, and other purposes was not explained in the captions.) In some cases neither the museum visitors nor even the curators knew the technological function of the objects. Therefore, although the artifacts were simultaneously both technology and art for their original users, they were not technology, but solely art, for the curators and viewers of the museum exhibit.

These examples suggest that for an artifact or piece of hardware to be technology, it needs to be set in the context of people who use it, maintain it, and repair it. This gives rise to the notion of a **technological system** that includes hardware as well as the human skills and organization that are needed to operate and maintain it (see consensus definition below).

Technology as Applied Science

Much of *contemporary* technology is applied science. However, to *define* technology simply as **applied science** is misleading both historically and systematically. If one understands science in the sense of the combination of controlled experiment with mathematical laws of nature, then science is only some four hundred years old. Even the ancient Greeks who had mathematical descriptions of nature and observation did not have controlled experiment. The medieval Chinese had highly developed technology (see chapter 10) and a rich fund of observation and theory about nature, but had neither the notion of laws of nature nor controlled experiment. Technology in some form or other goes back to the stone tools of the earliest humans millions of years ago. Clearly, with this understanding of science and technology, through most of human history, technology was not applied science. Part of the issue is how broadly one defines science. If one means by science simply trial and error (as some pragmatists and generalizers of Popper's notion of conjecture and refutation have claimed; Campbell, 1974), then prehistoric technology could be treated as applied science. However, now the notion of science has been tremendously broadened to include virtually all human learning, indeed all animal learning, if one holds a trial and error theory of learning. Perhaps this is an example of a definition of science that is too broad.

Even after the rise of early modern experimental science and the notion of scientific laws in the seventeenth century, and the development of the technology that contributed to the industrial revolution, most technological development did not arise from the direct application of the science of Galileo (1564–1642) and Newton (1642–1727). The inventors of the seventeenth and eighteenth centuries usually did not know the theories of mathematical physics of their day, but were tinkerers and practical people who found solutions to practical problems without using the science of their day. Even as late as Thomas Edison (1847–1931) we find a tremendously productive inventor in the field of electricity who did not know the electromagnetic theory of James Clerk Maxwell (1831–79) and his followers, but who produced far more inventions than those scientists who did know the most advanced electrical field theories. Edison initially even disparaged the need for a physicist as part of his First World War team, thinking one needed a physicist only to do complicated numerical computations, but that a physicist would have nothing much to contribute to technology. By this time Edison's view of the role of theory was getting somewhat dated.

Even in the contemporary situation, in which scientific training is essential for most technological invention, the notion of technology as applied science, if taken in too simple and straightforward a way, is misleading. Modern technology is pursued primarily by those with a scientific background and within the framework of modern science, but many of the specific inventions are products of chance or of trial and error, not a direct application of scientific theory to achieve a pre-assumed goal. Many chemical discoveries have been results of accidents. Safety glass was discovered when a chemical solution was spilled on a piece of glass laboratory apparatus, the glass was accidentally dropped, and it did not break. Penicillin was discovered when a bacterial culture was accidentally contaminated by a mold. Paper chromatography was discovered when a scientist accidentally spilled some chemical on a filter paper, and the chemical separated into two components as it seeped up the paper. The Post-it was discovered when a technologist, Art Fry, using little bookmarks in his hymnal, remembered a temporary glue that a colleague, Spencer Silver, had developed back in 1968 that was too weak to permanently stick two pieces of paper together. In 1977–9 3M began to market the invention, and by 1980 it was sold throughout the USA. Charles Goodyear's development of vulcanization of rubber involved numerous trials and experiments, but one crucial event involved him accidentally leaving his treated "gum elastic" on a hot stove, and noticing that it charred like leather. He then experimented to find a lesser, but optimum, heat of exposure (Goodyear, 1855). Louis Pasteur (1822–95) famously said that chance favors the prepared mind. The development of these accidental discoveries made much use of the scientific knowledge of the people who made them. But the discoveries were hardly the straightforward application of scientific theory to a preset problem.

For these reasons, although technology involves knowledge, particularly know-how, a definition of technology that characterizes it simply as applied science is too narrow.

Systems Definition as a Consensus Definition of Technology

A number of writers have formulated a somewhat complex definition of technology to incorporate the notion of a technological system. The economist John Kenneth Galbraith (1908–2004) defined technology as "the systematic application of scientific or other knowledge to practical tasks" (Galbraith, 1967, chapter 2). Galbraith describes this as incorporating social organizations and value systems. Others have extended this definition to mention the organizational aspect of technology, characterizing technology as "any systematized practical knowledge, based on experimentation and/or scientific theory, which enhances the capacity of society to produce goods and services, and which is embodied in productive skills, organization and machinery" (Gendron, 1977, p. 23), or "the application of scientific or other knowledge to practical tasks by ordered systems that involve people and organizations, living things, and machines" (Pacey, 1983, p. 6). We can combine these definitions into "the application of scientific or other knowledge to practical tasks by ordered systems that involve people and organizations, productive skills, living things, and machines."

This consensus definition is sometimes characterized as the "**technological systems**" approach to technology. The technological system is the complex of hardware (possibly plants and animals), knowledge, inventors, operators, repair people, consumers, marketers, advertisers, government administrators, and others involved in a technology. The technological systems approach is more comprehensive than either the tools/hardware or the rules/software approach, as it encompasses both (Kline, 1985).

WHAT IS TECHNOLOGY?

The tool approach to technology tends to make technology appear **neutral**. It is neither good nor bad. It can be used, misused, or refused. The hammer can be used to drive a nail or smash a skull. The tool user is outside of the tool (as in the case of carpenters' tools) and controls it. The systems approach to technology makes technology encompass the humans, whether consumers, workers, or others. The individual is not outside the system, but inside the system. When one includes advertising, propaganda, government administration, and all the rest, it is easier to see how the technological system can control the individual, rather than the other way round, as in the case of simple tools.

The notion (known as autonomous technology) that technology is out of human control and has a life of its own (see chapter 7) makes much more sense with technological systems than it does with tools. Technological systems that include advertising, propaganda, and government enforcement can persuade, seduce, or force users to accept them.

As noted above, not all students of technology wish to develop a definition or general characterization of technology. Some, particularly among the "postmodern" devotees of science and technology studies, claim not only that there is no "essence" of technology of the sort that mid-twentieth-century thinkers such as Martin Heidegger, Jacques Ellul and others claimed or sought, but that no general definition of technology is possible.

Despite the validity of the doubts of postmodern students of technology studies concerning an essence of technology, the "consensus definition" delineated above will help to keep the reader roughly focused on the kinds of things under discussion. For instance, the recent advocates of "actor-network theory" (see chapter 12) developed an approach to technology that has many affinities to the consensus definition in the technological systems approach. Advocates of the technological systems approach have recently begun to ally with or even fuse with the social construction of technology approach. Understanding technology as a network fits well with the European sociology of actor-network theory (see box 12.2). Thomas P. Hughes, the person who is perhaps the leading American historian of technological systems, has moved toward the social construction view, and combined it with his own approach (Bijker et al., 1987; Hughes, 2004).

Study questions

1 Do you think we can have successful discussions of controversial topics without bothering at all about definitions of major terms?

2   Can we make words mean anything we wish? In what sense is this true and in what sense is this false?

3   Are there any areas of knowledge or subject matters in which there are "real definitions"? Are there any areas in which there are essential definitions?

4   What sorts of classes of things might have only "family resemblances" and no essential definitions? (Give examples other than the two given in the chapter and explain your answer.)

5   The philosopher Arne Naess (decades later to be the founder of "deep ecology") in his earliest work surveyed people on the street as to their definitions of various philosophical terms. What sort of definitions was he collecting? Do you think this is a fruitful way to clarify philosophical issues?

6   Do you think that the characterization of technology as applied science is correct? Give examples that support this characterization and examples that go against it (other than ones given in the chapter).

7   Does the notion of technology without tools make sense? If not, why not? If so, try to give some examples not mentioned in the chapter.

3

Technocracy

Technocracy is a theory of rule by technical experts. (Various other similar terms for rule are: democracy, rule by the *demos* or common people; aristocracy, rule by the *aristos* or best; and plutocracy, rule by plutocrats or the wealthy.) Theories of technocracy have differed over exactly which sorts of experts are fitted for rule, ranging from pure scientific or engineering expertise to including the social scientific expertise of economists and sociologists. In various forms, blatant and subtle, technocratic notions have been present in the attitudes of many twentieth-century and present policies. The extension of the prestige and authority of technical experts to authority in non-technical, especially political and economic, decision-making is an implicit technocratic development.

This chapter surveys the major figures who have advocated rule by an intellectual or technical elite. It turns out that they also comprise most of the major historical philosophers who have written on issues related to technology. As noted in the Introduction, not many of the major past philosophers dealt at length with technology. However, Bacon around 1600, as well as St Simon and Comte in the early nineteenth century, are three of the early major philosophers of technocracy, and are also advocates of technocracy in one form or another. Plato, who is the first Western philosopher by whom we have extensive writings, is also one of the towering figures of Western philosophy. The twentieth-century philosopher and mathematician Alfred North Whitehead was hardly exaggerating when he said, "The safest general characterization of the European philosophical tradition is that it consists of a series of footnotes to Plato" (Whitehead, 1927, p. 39). He also was the major inspiration for later utopias and for theories of rule by an intellectual elite. We re-encounter Bacon, who was a major precursor of technocratic thinking as well as a scientific methodologist. St Simon in France in the early nineteenth century proposed fully fledged technocratic theories, with physical and social scientists, respectively, as the rulers. Then we examine early twentieth-century technocracy, whose major thinker was the American economist and sociologist Veblen. Finally, we survey late twentieth-century technocratic thinkers, who had influence on government and policy, including "postindustrial" social theory.

Plato

The word "technocracy" dates from the 1920s, but the roots of the notion of technocracy go far back into Western history. In ancient Greece Plato (*c*.428–347 BCE) proposed rule by philosophers in his *Republic* (*c*.370 BCE). However, Plato's proposed training of his philosopher kings included a great deal of advanced mathematical education. The reason for this was that Plato believed there were real structures and natures of things that he called the forms (see the section on "real definitions" in chapter 2). Plato argued that there are ideal forms of shapes (geometrical forms) and of physical objects. There are also ideal forms of ethical notions such as courage, piety, and justice. These could be known only by purely intellectual grasp, not by sense perception, which Plato considered a lower and less accurate form of knowledge.

In the famous Allegory of the Cave in the *Republic* Book VII, Plato compares ordinary humans to prisoners chained in a cave who are entertained by the shadows on the wall of puppets. They see only the shadows and neither the puppets that cast the shadows nor the fire that illumines the wall (it has been suggested that Plato may have used the technology of puppets of his day as the basis of his metaphor for the illusion of common-sense knowledge; Brumbaugh, 1966). Plato tells the tale of a person who descends into the cave and frees one of the prisoners to ascend and be exposed to direct sight of physical objects and finally to brief glimpses of the sun. Plato claims that ordinary humans know only the shadows of the forms as physical objects. Intellectual education can lead individuals to grasp the forms and eventually to glimpse the Form of the Good. Plato's scheme of education for the rulers of his Republic is this journey to the light.

Plato not only presented his ideal Republic as ruled by an elite educated in the highest forms of reasoning, but, quite unsuccessfully, attempted to persuade the tyrant of Sicily to institute some of his ideas. According to one story, Plato was sold into slavery by the angry ruler, and his wealthy friends had to pay ransom for him. (The dramatic story of Plato's expedition to Sicily is told in his Seventh Letter, a work that Plato ought to have written, whether or not he is the real author.)

For the purposes of political rule Plato was interested in the **forms** of justice and other ethical notions. Mathematics presented the clearest example of precise intellectual knowledge of the forms (of numbers and geometry). However, mathematics also exemplified strict reasoning from determined assumptions. Mathematics remained the model of all rationality for many later philosophers (see chapter 4). The geometry of Euclid (c.365–275 BCE) begins with a set of assumptions, axioms, and postulates, and by logical steps deduces the results of geometry. In the educational program for rulers in Plato's *Republic* mathematical study was only a preliminary to higher philo-sophical study as a preparation for rule. While soldiers and craftspeople needed very simple geometry and arithmetic for tactics or trade, the rulers were trained for a decade in the theoretical mathematics of the day, includ-ing a purely mathematical astronomy and theory of music. Plato even claimed that pursuers of this pure mathematics should not concern themselves with astronomical phenomena or hear music. Once the advanced mathematics of the day had been mastered, the rulers were introduced to philosophical reasoning or dialectic. Plato claimed that mathematical reasoning started with basic assumptions or axioms. Philosophical dialectic was a higher form of knowledge, in that it questioned and evaluated the basic assumptions of knowledge. Thus Plato's rulers or philosopher kings and queens were not genuine technocrats. Their mathematical training was only a prelude to acquiring philosophical wisdom and the ability to reason about moral and political matters.

In much of the later Platonic tradition the sharp distinction between math-ematics and the higher realm of philosophy was blurred or erased. Some later neo-Platonists (and even, allegedly, Plato himself, according to contro-versial reports of his later, unwritten teachings and of the "Lecture on the Good" allegedly delivered at the Academy) shifted the emphasis to math-ematical knowledge or mathematics-like knowledge as the key to all philo-sophy. Plato's nephew Speusippus (410–337 BCE) took over the academy of Plato and replaced the forms with numbers. Plato's greatest student; Aris-totle, claimed that the immediate successors of Plato mistakenly identified philosophy with mathematics. Thus the line between mathematics and philosophy was partially erased in much of the later neo-Platonic tradition, opening the way to universalizing the mathematical model of knowledge in later technocratic thought. The power, rigor, and prestige of mathematical reasoning plays an important role in modern technocratic thought, insofar as engineers, economists, and others who use mathematical methods can associate that aura of precision and rigor with their pronouncements in politics and other areas beyond their specialties.

Francis Bacon

Francis Bacon (1561–1626) in the early modern period presented a utopia much closer to a genuine technocracy than Plato's *Republic*. Bacon was an extraordinary figure of the English Renaissance. He modestly claimed to take "all knowledge for his province," and did so. He became Lord Chancellor of England and is famous as the author of succinct and pithy essays. (Some have so admired his writing style as to claim, most implausibly, that he was the author of Shakespeare's works.) Bacon was active in law, philosophy, and science. After working his way from a penniless youth to a figure of great wealth and power he was indicted for accepting a bribe (although this was common practice at the time and his indictment was likely a product of political conflicts). His death, according to legend, was caused by bronchitis brought on from getting a chill while experimenting with preserving a chicken by stuffing it with snow. In order to maintain his status he chose for recuperation a wet and cold bed in an elegant room in a castle over a warm, dry bed in a much smaller room, and this choice of aristocratic elegance over efficiency killed him.

Francis Bacon believed that through both knowledge of nature and technological power over nature humans can regain the clarity of mind and purity of action that Adam and Eve had before the expulsion from the Garden of Eden. Despite this religious formulation of the goal, Bacon in his *New Atlantis* described an ideal society in which pursuers of something closer to the modern notion of scientific and engineering knowledge played a central role in the running of a prosperous and healthy society.

In Bacon's (1624) utopia, *New Atlantis*, Salomon's House was the name for a kind of national research institution. Here experiments were carried out and the properties of minerals, plants, machines, and many other things for the purpose of the improvement of life were studied. Salomon's House even included individuals who worked as what we today call industrial spies, traveling in disguise to other countries to observe their crafts and manufactures. Bacon unsuccessfully proposed a more modest but real version of

41

Salomon's House in the form of an actual college that would contain zoological and botanical gardens, laboratories, and machine shops. Unfortunately, King James I, sponsor of the King James Version of the Bible, showed no interest in such an enterprise, or in Bacon's dream of social betterment for the nation.

Bacon also advocated experimental science and the **inductive method** (see chapter 1). Theories of nature, he argued, should be based on generalization from individual observations and tested by individual observations, not deduced from general principles. Bacon's empirical approach was the opposite of Plato's. Where Plato downgraded the status of sense perception to veritable illusion and claimed that pure intellectual reasoning was the way to truth, Bacon claimed that sense observation was the road to truth and that theories spun out of pure reason and philosophical speculation were worthless paths to error. Bacon presented his theory of the sources of error in his *New Organon* (1620) (modestly titled to suggest that it was replacement for the two millennia old *Organon*, or logical works of Aristotle). In evaluating our observations Bacon claimed that we constantly had to be on guard against what he called "the idols," or distortions of perception and thought, to which humans are prone and which they need to correct. He classified these as: (a) the Idols of the Tribe, which are features of the human constitution that mislead us, such as perceptual illusions and biases, as well as cases where our hopes and preconceptions distort our perceptions; (b) the Idols of the Cave (referring to Plato's cave), which are illusions particular to the subjective experience and background of the individual; (c) the Idols of the Market Place, which are products of human communication, particularly the distortions and ambiguities of language; and (d) the Idols of the Theater, or delusions produced by belief in speculative philosophical systems.

Bacon's idea of empirical scientific research inspired some decades later a number of founders of the Royal Society, the premier British scientific society, which exists to this day. Just as Plato taught that the intellectual of the forms, most evident in mathematics, could be applied to ethics and politics, Bacon believed that his inductive method should be applied to jurisprudence, deriving legal maxims by induction from legal cases the way one derives scientific laws by induction from particular observations. Bacon stated that knowledge is power and that one could command nature (control it) by obeying it (grasping the principles and causes). Bacon often analogized the relationship of the inquirer to nature as that of man to woman and used metaphors of seduction, unveiling, and force to describe the process of inquiry (see chapter 9). Bacon's claim that knowledge is power and that investigation of nature is the way to social prosperity and well-being is much closer to the technocratic notion of Plato's philosopher-rulers. However, the investigators at Bacon's Salomon's House did not themselves seek to rule. They merely advised the rulers (prudently not specified or described in his *New Atlantis*).

Claude Henri de Rouvroy, Comte de St Simon

While inquirers at Salomon's House did not themselves directly rule society, in some of the schemes of the French Comte de St Simon (1760–1825) in the early nineteenth century, scientists and technologists did directly rule society.

St Simon participated in the American Revolution, offered his services elsewhere as a soldier of fortune, and supported the French Revolution, renouncing his title and presiding over a meeting of his community. In the wake of the French Revolution he made a fortune speculating on the property abandoned by fleeing royalists and on the deserted churches that had been closed by the revolutionaries. He made money by selling the lead from the windows and roofs of churches, attempting at one point to sell the roof of Notre Dame in Paris. According to legend his valet was under orders to awaken him each morning by announcing, "Arise Count, you have great things to do today!" He allegedly proposed, unsuccessfully, to the leading literary woman of the day, Madame de Staël, by saying "Madame, you are the most extraordinary woman on earth and I am the most extraordinary man; together we shall produce an even more extraordinary child" (Heilbroner, 1953).

St Simon, no technical man himself, and largely self-educated (by inviting the leading scientists of the day for dinner conversations), gathered a circle of engineering students from the leading French technical school of the day, *L'Ecole Polytechnique*, founded during the French Revolution and supported and further developed by Napoleon. St Simon saw the older feudal society as wasteful, superstitious and warlike, supporting numerous parasites in the form of the nobility and the clergy. He provocatively opens one essay with the claim that if one morning the nation woke up to find all the clergy and nobility gone the nation would not suffer, but if all the scientists, technicians, and business people were gone society would collapse (St Simon, 1952, p. 72). The new alternative that St Simon saw emerging was industrial society.

During the course of his lifetime, St Simon sketched a number of different plans for the rule of industrial society. They varied with the tumultuous politics of the times, whether dominated by radical revolutionaries, royalists, or businessmen, probably with an eye to support from the rulers or the powerful of the society of the day. Some were radically socialistic, and some were capitalist, but all were quite centralized (like later French capitalism). St Simon called the ruling body of his ideal society the Council of Newton, suggesting the scientific nature of the rule by associating it with the name of the pre-eminent physicist of the previous centuries. Scientists, technicians, industrialists, and bankers filled the seats of various versions of the council. Priests and nobility were eliminated from the new society.

St Simon's ideas were influential on politically divergent groups. Some of the early French capitalist proponents of railroad lines and of what was to become the Suez Canal were followers of St Simon, but so were various socialist revolutionaries (Manuel, 1962, chapter 4). St Simon coined a number terms that came into worldwide use, including "individualism," "physicist," "organizer," "industrialist," and "positivist" (Hayek 1952). The word "socialism" was not coined by St Simon, but soon appears in a St Simonian journal written by his followers. Both the capitalism and the socialism that St Simon variously advocated were planned and centralized. His capitalism was one controlled by banks and monopolies. St Simon's technocratic and centrally planned version of socialism resembles that of the USSR. A number of his slogans found their way into the Communism of Lenin and Stalin in the Soviet Union via the writings of Marx's collaborator Friedrich Engels (1820–95). (Engels introduced a number of St Simonian phrases and ideas that were not present in Marx's draft of *The Communist Manifesto*.) Lenin described the future social organization of communism using the St Simonian terms "society as one vast factory," and "the organization of things and not of men." Stalin used St Simon's phrase to characterize "artists as engineers of the human spirit." Ironically, a similar notion appears in the designation of the technicians at Disneyland as "imagineers" (or engineers of the imagination). St Simon's views, unlike those of Plato and Bacon, present a fully fledged technocracy. In some of his presentations, at least, the experts literally rule.

Auguste Comte

Auguste Comte (1798–1857) attended the *L'Ecole Polytechnique* and studied the physical sciences of his day, but was expelled after being involved in a protest against the takeover of the school by the new monarchy that followed the defeat of Napoleon. Comte began as a follower of and assistant to St Simon. Comte systematized and greatly expanded the scattered and disorganized ideas and lectures of his teacher St Simon. One of Comte's most central doctrines was his **Law of the Three Stages**, which claimed that society evolved from a theological or religious stage, through a metaphysical or philosophical stage, to a final positive or scientific stage. In the **theological stage** the reasons for things are believed to be due to a will or wills. At first, in fetishism, every object has its own will. Then, in polytheism, there are the wills of a number of gods. Finally, in monotheism, a single divine will accounts for everything. In the **metaphysical stage** the causes of things are considered real abstractions, such as powers and forces. Finally, in the **positive stage** the search for ultimate causes is relinquished and laws of succession are the goal of knowledge. For Comte religion and metaphysics are inferior, less evolved forms of knowledge in comparison to scientific knowledge. The superior status of scientific knowledge and the downgrading or dismissing of other, non-scientific forms of knowledge in the humanities is very much part of the technocratic creed.

The first part of Comte's philosophy is a theory of scientific knowledge and philosophy of science. The second part is a social and political philosophy of the organization of industrial society. Comte associated the forms of thought of the three stages with forms of society. The theological stage of society is militaristic. The metaphysical stage of society is centered on law and jurisprudence. Finally, the positive stage corresponds to industrial society.

Comte's later work on social organization and the "religion of humanity". may have been influenced by his love life, whether in a new appreciation of the emotions in his love for Clothilde de Vaux, or in guilt over his mistreatment of his companion. In Comte's industrial society scientists *literally* replaced the Roman Catholic priesthood. A hierarchy of technical experts replaced the church hierarchy.

In the twentieth century, books about the scientific elite had titles, such as *The New Priesthood* (Lapp, 1965), that were meant metaphorically, but Comte meant this quite literally. Comte's plan for positivist churches that were to replace the Catholic Church was a non-starter, although a few existed from Brazil to England. The flag of Brazil bears Comte's slogan "Order and Progress." Influential figures in Mexico during the late nineteenth-century rule of Porfirio Diaz (1830–1915) professed allegiance to positivist ideals (Zea, 1944, 1949).

Comte's positivism was highly influential in less explicit and blatant forms throughout the twentieth century. Rule of society based on knowledge means rule of society based on scientific knowledge. Politics becomes a form of applied science or social engineering. Comte invented the field and coined the term "**sociology**" for the scientific study of society. He saw sociology as the master discipline of social rule. For Comte, sociology, though not at the basis of the hierarchy of science as was physics, was truly Queen of the Sciences. Thus Comtean technocracy included social scientists, not just physical scientists, in a central role in the rule of society, as was often the case in St Simon. Technocracy as rule, not specifically by engineers but by social scientists of various sorts, became characteristic of the theories of a number of social theorists in the USA and Europe in the 1950s through 1970s (see below).

Comte's philosophy of science and philosophy of history (the three stages) were more long lasting in their direct influence than Comte's plans for social organization and the "religion of humanity." Logical positivism (see chapter 1) eliminated Comte's political and religious theories. Nevertheless, logical positivism retained an updated form of Comte's claim that science is the highest form of knowledge, indeed the only genuine knowledge.

Thorstein Veblen and the Technocracy Movement in the USA and Elsewhere

During the early twentieth century the term "**technocracy**" was introduced for the first time. The economist John M. Clark coined the term "technocracy" in the mid-1920s. In the USA there was an actual technocratic movement named as such in politics. Its heyday was in the 1920s and 1930s and the movement was always small. It still exists today but is hardly noticed.

The economist Thorstein Veblen (1857–1929) was the major American theorist of technocracy in the early twentieth century. Veblen was a detached and arch critic of the Gilded Age society of late nineteenth- and early twentieth-century America. He studied the customs of the business elite of his day the way an anthropologist studied an exotic pre-civilized culture. He despised the prissy academic culture in which he taught, and his book about universities, *The Higher Learning in America*, was originally to be subtitled *A Study in Total Depravity* (Dowd, 1964). He had almost equal disdain for economic formalism. When *The Theory of Business Enterprise* was refused by a publisher for not having enough mathematical economics in it, he simply added some fake equations in footnotes. Veblen's irreverent and iconoclastic remarks and his risqué lifestyle caused him to be separated from several of the major universities at which he taught (at one point he lived in a tent in a pasture just outside the perimeter of the Stanford campus). He is probably most famous for his notion of "conspicuous consumption" in *The Theory of the Leisure Class* (1899), in which ostentatious display of one's purchases is utilized both to advertise one's wealth and importance and to intimidate the lower classes.

In *The Engineers and the Price System* (1921) and other works Veblen contrasted the wastefulness and inefficiency of business practices with the efficiency of the engineers. He contrasted the human "instinct for workmanship" manifested in modern times most fully by the engineer with the predatory instincts of the businessman. He proposed a society run by engineers rather than by businesspeople. During the economic downturn after the First World War and in the wake of the Russian Revolution Veblen proposed a genuine revolution of the engineers and even spoke somewhat facetiously of "a Soviet of engineers" (though he soon backed away from a direct proposal of political action).

A follower of Veblen and self-proclaimed engineer, Howard Scott continued the political technocratic movement (Scott hung out in Greenwich Village coffee houses in New York and was a kind of coffee house engineer analogous to the coffee house poets of the period). Scott's activism and following was revived during the Great Depression of the early 1930s. Scott's technocrats combined the conditioning of human behavior, following the behaviorist psychologists Ivan Pavlov (1849–1936) and John Watson (1878–1958), with their conception of society run as a machine by engineer experts. The political technocratic movement in the USA, like Comte's positivist churches, did not ultimately catch on. Scott's Technocracy, Inc., with its uniforms, identical automobiles, and yin–yang symbol icon, was widely discredited after the failure of a much-anticipated national radio address by Scott (Elsner, 1967). By 1936 the political movement that explicitly called itself technocracy lost its popular appeal but it exists to this day as a small sect that runs advertisements in liberal and leftist magazines.

The technocratic tendency diffused widely throughout American society in the Progressive movement of the period around the First World War and in the New Deal of President Franklin Delano Roosevelt (1882–1945) during the 1930s. Both the Progressive movement and the New Deal were responses to social disorganization and crisis. The Progressives reacted against corrupt city political machines and price-gouging monopolies. The New Deal was a response to the Great Depression (1929–1941). However, the phrase social engineering had already become current among politicians of the Progress-ive Movement and among followers of American pragmatism in politics. Even pro-capitalist engineers such as Herbert Hoover (1874–1964), later to be US President, proclaimed the ideal of the engineer as manager of an efficient society. Nonetheless, all but a handful of engineers drew back from Scott's radical conclusion of the promotion of engineers to be rulers.

Technocracy was not limited to the USA. In Sweden, the early works of now famous sociologists Gunnar and Alva Myrdal advocated technocracy. The Myrdals, however, unlike many other technocrats, clearly saw the dan-gers of the conflict of technocracy with democracy, and attempted to syn-thesize the two (Myrdal, 1942). In Germany, Karl Mannheim, the sociologist of knowledge, wrote on "democratic planning" and on the role of the elite of free-floating intellectuals in the 1920s, and continued writing on this theme after he fled to Britain in the 1930s (Mannheim, 1935; 1950). Mannheim's and the Myrdals' technocratic social planners were social scientists, not phys-ical technologists.

Meanwhile, both totalitarian dictatorships of the period, Germany and the USSR, had strong technocratic components. Nazism incorporated a strange mixture of anti-technological, pagan, health food, nudist, back-to-nature rhet-oric, with a technocratic belief in the capacity of engineers of new technology to bring the regime to world power. During the war the technocratic aspect won out over the romantic, ecological element (Herf, 1984; Harrington, 1996, pp. 193–9). In the USSR a strong technocratic ideology was present in the Stalinist rhetoric of forced industrialization (Bailes, 1978). As noted, the central planning ideal of the USSR resembled St Simon's dream.

Technocracy and Post-industrial Society Theory

During the 1950s, 1960s, and 1970s in the USA, the welfare states of Europe, and the communist USSR, technocratic tendencies were influential in theor-ies of government. During the New Frontier of the presidency (1961–3) of John F. Kennedy and the Great Society (1963–8) of Lyndon Johnson in the USA, and in the Labor government (1964–70, 1974–6) of Harold Wilson in the UK, technocratic notions were afloat among advisors. Some spoke of the "white-hot" technological revolution, and Wilson in a preface to a Labour Party report claims his party "believed that manpower and resources must be planned intelligently . . . [but] in the modern world such planning would be meaningless without the full planning and mobilization of scientific resources" (Werskey, 1978, p. 320). In the USA, former General Motors executive Robert S. McNamara and his "whiz kids" were the experts in quantitative analysis who did the strategic planning for the Vietnam War (1961–73). After the Second World War the centrality of scientific research and development to the economy was recognized. Big Science, such as the atomic bomb project at Los Alamos, had come of age during the Second World War. Nuclear physicists, such as, successively, J. Robert Oppenheimer (1904–67) and Edward Teller (1908–2003) (Herken, 2002), as well as the mathematician John von Neumann (1894–1964) (Heims, 1980; Poundstone, 1992), became important advisors to government during the Cold War nuclear arms race. President Dwight Eisenhower, in his in 1960 farewell address, famously warned of the growing power of the "military industrial complex," the complex of large corporations producing military armaments and the Pentagon bureaucracy.

In the USA and in Germany it was claimed by various sociologists that political ideology had become irrelevant and what was important was tinkering to fine tune the economy by economists and social planning by technocratic social science experts (Bell, 1960; Aron, 1962; Dahrendorf, 1965). This was the "end of ideology" thesis. In contrast, in the Soviet Union, Marxist-Leninist ideology was claimed to be the true theory of politics. In the Soviet Union Marxism-Leninism played the role of the technocratic social science of central planning. It was claimed to be the science of society (and even of nature) on the basis of which political decisions were made. In Western Europe ideology had a less pejorative connotation than in the USA, and by the end of the century often had a positive one. In the USA, political commentators still discussed "ideologists" versus "pragmatists" as the factions in the People's Republic of China.

In the West, the theory of **post-industrial society** was advocated by a number of technocratic thinkers in the 1960s and 1970s. These thinkers included the economist John Kenneth Galbraith (1967), the sociologist Daniel Bell (1973), and the foreign policy advisor Zbigniew Brzezinski (1970). Post-industrial society theory is a kind of technological determinism (see chapter 6). It claims that various forms of the technology of industrial production produce different forms of social rule. In this respect it is like Marxism, but it rejects Marx's socialism and communism for a prediction of the coming of technocratic rule in post-industrial society.

Post-industrial society theory describes the stages of society as agricultural followed by industrial followed by post-industrial. Agricultural production using human and animal power along with some wind and waterwheel power yields a society of peasants and feudal rule. Mechanized manufacturing industry leads to blue-collar factory workers and rule by capitalist owner-entrepreneurs. Finally, the rising dominance of information processing and service industries leads to new forms of educated workers to oversee automated machinery and also leads to technocratic rule. In industrial society agricultural workers, once the vast majority of the population, become a small minority. So, it is claimed, in post-industrial society blue-collar factory workers shrink to a minority. Information technology rather than energy technology becomes dominant. Brzezinski even claimed that the student revolts of the 1960s were similar to the peasant revolts of the early modern period, in that the desperate, rebelling humanities students found themselves to be superfluous in a society that was to be run by computer scientists and engineers.

Large corporations owned by numerous stockholders but run by managers replace the traditional, family owned and run firms of earlier capitalism. Post-industrial society theorists claim that the new society is post-capitalist in that the capitalists, the owners of stock, no longer run the firms. Instead, a variety of planners, engineers, industrial psychologists, advertising, marketing and media experts, economists, and accountants supply the information to the managers. The "separation of ownership and control," first described by Franklin Roosevelt advisor Adolph Berle (Berle and Means, 1933) and later developed by economist John Kenneth Galbraith (1967), describes the situation. Galbraith and others went as far as to claim that long-term rational planning in the corporation under managers comes to dominate the traditional search for short-term profits by capitalists. Some conservative post-industrial society theorists claim that a "new class" is replacing capitalists as most influential in society. This class is variously identified with technocrats or with managers, and is sometimes described as the professional managerial class (PMC).

The technocracy thesis for post-industrial society has a simpler and a subtler form. The simpler form is that the stratum of technocratic experts, what Galbraith calls the **technostructure**, directly rule, replacing traditional politicians and business leaders. The subtler form, suggested by Galbraith (1967), is that politicians and corporate chief executives are dependent for information on the basis of which to make decisions on numerous lower level technical experts, scientists, engineers, accountants, economists, political scientists, psychologists of propaganda and the media, and so forth. These often invisible lower level figures frame and even bias the alternatives between which the politician or executive chooses, and thus surreptitiously channel and direct policy. So-called "policy wonks" and "computer nerds," to use the disparaging slang applied to them, actually control the direction of the state, despite their lack of visibility. This form of the technocracy thesis has relevance even when the national leaders profess political views that would reject technocracy, but nevertheless are dependent for their decision-making on numerous economists, military technology experts, political scientist pollsters, and science advisors. (For instance, 1988 presidential candidate Michael Dukakis was disparaged as a "technocrat," although the elder George Bush, as former head of the CIA, was hardly detached from the technostructure.)

Conclusion

Technocracy is a notion with a long prehistory and a variety of forms in contemporary society. Plato emphasized the knowledge in rule and used mathematics as a model of intellectual knowledge and a means of training rulers, although the rulers themselves were philosophers. Bacon emphasized the power of knowledge of nature and presented a utopia in which investigators of nature supplied information to the rulers and in which the exploitation of nature led to prosperity and power for the state. St Simon and Comte emphasized the superiority of scientific knowledge to religion and philosophy and directly advocated a ruling role for scientists and engineers. In the early twentieth-century USA, the word technocracy and an actual political technocracy movement arose. The movement had only a brief span of interest and popularity, but the technocratic notion became widespread in less blatant forms. In the USA, Western Europe, and the USSR during the 1950s through the 1970s more subtle forms of technocratic doctrine were widespread. In the USA and Western Europe political ideology was said to be obsolete and replaced by social and economic engineering during the late 1950s. In the 1970s, theorists of post-industrial society claimed that traditional owner-entrepreneurs and traditional politicians were being replaced by corporate and government technocrats in the information society.

Claims about the disappearance of political ideology were quickly shown to be false by the ideological political protest movements of the 1960s. The demise of the short-term profit-oriented owner was shown to be exaggerated in the 1980s and 1990s. Nevertheless, diluted or implicit technocratic

51

ideas have been widespread in government agencies and in social theory. In a society tremendously dependent on technology and technological development, with large corporations and government dependent on economists and other social scientists such as marketing psychologists, survey-takers, and pollsters, the technocratic tendencies of society continue even if they are widely condemned rather than praised. One of the issues raised by the technocracy thesis is whether the only form of rigorous and useful reasoning is scientific and technological reasoning, or whether there are non-technical forms of reasoning appropriate and applicable to social issues and the problems of everyday life. We discuss this issue in the next chapter.

Study questions

1   Is technocracy desirable? Why or why not?
2   Is the "subtle" version of technocracy, Galbraith's technostructure, true of your society? That is, do leading politicians and corporate leaders have their decisions pre-framed by technical advisors to the extent that their decisions are directed by the lower level input?
3   Does the shift of our economy from a manufacturing economy to a "service economy" justify the claim that it is becoming a technocracy? What sort of jobs count as "service" jobs? Are all or most of these relevant to the high technology economy?
4   Adolph Berle in the 1930s and J. K. Galbraith in the 1960s claimed that in the modern corporation there is a separation of ownership from control. That is, the corporations are owned by stockholders who do not oversee the day-to-day operations of the corporation. Management, which does not own the corporation, controls its operations. Do you think that this accurately describes today's corporations? Does it apply to all, some, or none of them?

4

Rationality, Technological Rationality, and Reason

An issue that divides boosters and detractors of technology and the technological society is the nature of rationality. Science is generally taken as the prime model or paradigm of rationality in our society. Technology (usually construed as applied science) is likewise seen as part of the rationality of modern society.

Technocrats see themselves as advocates of the rule of reason. Unlike Plato, however, they understand reason to mean technological/scientific reason. Analytical critics of the technological pessimists or dystopians distrust the grand theses of European figures such as Heidegger and Ellul. Technocrats and most analytical philosophers of technology advocate a piecemeal evaluation of technology, one project at a time (Pitt, 2000, chapters 5, 6). In this, ironically, they agree with the recent continental philosophy influenced American philosophers of technology (Ihde, Feenberg, Haraway). These philosophers are skeptical of the claim that technology has an essence or general character that can be morally or culturally assessed as a whole (Achterhuis, 2001, pp. 5–6). Here analytical philosophers and postmodernists (odd bedfellows, indeed) agree. Many analytical philosophers and almost all technocratic opponents of grand theses and narratives of technological pessimism generally use risk/benefit analysis to do the evaluation (see below). It is a major question whether the mathematical calculations of risk and benefit can incorporate, or do justice to, the moral and aesthetic values of the people who live with the technology, as we shall see toward the end of this chapter.

Many students of the rise of modern society, beginning with the early twentieth-century German sociologist Max Weber, have portrayed the rise of modern, Western society as the rise of rationality. Weber spoke of the "rationalization" of various areas of society, including, of course, economics and science, but extending through all areas of society and culture. By rationalization Weber meant systematization and organization by means of rational principles. Weber included in his extraordinarily wide-ranging study of rationalization not only bureaucracy, but also theology (in Judaism, Confucianism, Taoism, Buddhism, and Hinduism). He even included, as an extensive example of rationalization in music, the history of the development of the piano (Weber, 1914, 1920, 1920/1a, b, c).

Jacques Ellul's "technique" has many parallels to Weber's "rationalization." (Oddly, Ellul in *The Technological Society* (1954), his first and most famous work, which introduces the notion of technique, does not mention this concept of Weber.) Recall that Ellul is a prime advocate of the notion of technology as primarily a matter of rules rather than of hardware (see chapter 2). Technological rules constitute his "technique." Ellul's "technical phenomenon" is the application of technique to all aspects of life and society, and corresponds to the complete triumph of Weber's process of rationalization.

In late twentieth-century theories of technocracy and post-industrial society the application of scientific rationality to various areas of social prediction and planning was seen as an admirable culmination of the rise of reason. The application of such techniques as operations analysis, cost/benefit and risk/benefit analysis, rational choice theory, and the general application of economic models to apparently non-economic aspects of society, such as politics, and even mate choice, is seen as a positive step. Applied social science becomes "social engineering" of a sort far more complex and sophisticated than the Progressive Movement forerunners of the technocracy movement envisaged (see chapter 3).

In contrast to the technocrats and technological optimists, those who have been pessimistic about the dominance of technology in our society have often contrasted a higher or genuine rationality with **technological rationality**, or "**instrumental rationality**" (see below). Technological rationality is seen as a lower form of rationality that needs to be supplemented and overseen by genuine philosophical, dialectical, or other higher rationality. This is particularly true in the German tradition that comes from Immanuel Kant (1724–1804) and Georg Friedrich Hegel (1770–1831) in the late eighteenth and early nineteenth centuries. This contrast of dialectical and instrumental reason is taken up in twentieth-century critical theory.

One traditional model for rationality in the West since the time of Plato in ancient Greece has been mathematics (see chapter 3). Mathematics is generally considered to have the features of universality, necessity, rigor, and certainty. Mathematics has universality with respect to individuals as well as cultures. Mathematical results are such that anyone correctly following the calculation techniques will get the same result. There is no subjective individual variation in correct answers to a well set problem. Likewise, there is no cultural variation in the results of a proof or a problem even if there is cultural variation in notation or symbols. The "Pythagorean theorem" (concerning the squares of the sides of a triangle containing a right angle) was independently discovered in the ancient Near East and in China, but the result is not relative to the different cultures that have discovered and used it (some students of ethno-mathematics will dispute the generalization of this claim; see chapter 10). The logic of mathematical proof has a compelling necessity. If one follows the proof, step-by-step, one is inexorably led to the conclusion. The necessary conclusions have certainty. They cannot reasonably be doubted. Mathematical algorithms show this necessity and certainty with particularly clarity. An algorithm is a procedure that gives rules that if followed will lead mechanically to the correct result. Mathematical results are precise and not vague. Even mathematics that deals in probabilities and statistics gives precise probabilities and distributions.

These features of mathematics have led many philosophers and theorists of society to see mathematics as the paradigm of rationality. Many Western philosophers have thought that rationality in general should aspire to the universality, necessity, certainty, and precision believed to be exhibited by mathematics. The seventeenth-century philosophical movement now known as rationalism, whose main representatives were the French mathematician-philosopher René Descartes (1596–1650), inventor of analytic geometry, the German mathematician-philosopher Gottfried Leibniz (1646–1716), co-inventor of the differential and integral calculus, and the Dutch philosopher Baruch Spinoza (1632–77), who worked as a lens grinder to support himself, aspired to make all philosophical reasoning conform to the necessity and rigor of mathematics. Spinoza cast his book *Ethics* in the logical form of a geometry treatise, with axioms, theorems, and proofs. Even philosophers who believed that our reasoning in science and in ethics fell short of the mathematical ideal used the mathematical ideal as the standard by which to measure reasoning in other fields. John Locke (1632–1704) claimed that natural philosophy (physics) could not be a science because we do not know the essences of the submicroscopic construction of bodies (1689, p. 645), while ethics might be a science because it is based on logical derivations from definitions! Although Locke is often considered the founder of empiricism, this evaluation of physics and ethics is the opposite of that of later logical positivism and much contemporary educated (or half-educated) common sense.

This mathematical ideal of reasoning has led in recent centuries to computational models in ethics. The British philosopher Jeremy Bentham (1748–1832) claimed that ethics was a matter of adding units of pleasure and subtracting units of pain (as negative of pleasure). Actions and policies that maximized pleasure and minimized pain for all concerned (computed this way be simple arithmetic) were the best acts and policies. Bentham called this theory of ethics **utilitarianism**. Rational choice theorists in twentieth-century political science have modeled political and military strategies on an economic model of costs and benefits. **Risk/benefit analysts** evaluate the worthwhileness of technological projects by adding benefits and subtracting risks in a manner similar to Bentham's utilitarianism.

Scientific rationality is broader than mathematical rationality. Science involves mathematics, but it also involves observation and experiment. The support or confirmation of scientific hypotheses and theories by evidence does not involve the necessity and certainty of a mathematical proof or algorithm. Scientific hypotheses and theories are not certain, but are, at best, probable. Even the estimation of the probability or degree of support or justified belief in scientific statements is not mechanical. Science involves guesswork and judgment.

Nevertheless, many philosophers of science, inductive logicians, and logical positivists during the nineteenth and much of the twentieth centuries pursued a mechanical version of the scientific method and an algorithm for automatically and exactly computing the probabilities of scientific theories. This was the ideal of Rudolf Carnap's work in formal inductive logic. All but a very few philosophers of science by the later decades of the twentieth century concluded from the failure of this enterprise that this was an illusory goal, and that scientific method cannot be made mechanical and algorithmic.

A number of thinkers in recent decades, particularly philosophers of science, persuaded of the failure of algorithmic models of science, particularly the failure of the program for a formal inductive logic, have embraced a wider conception of rationality that involves judgment (Putnam, 1981, 174–200; Brown, 1988). Judgment involves the sizing up of a situation, assessment of evidence, and deciding on a path of action without following rules. Aristotle, in his *Ethics* (c.240 BCE), emphasized the role of judgment. Immanuel Kant, especially in his Third Critique, *The Critique of Judgment* (1791), also emphasized the role of the power of judgment, and claimed that judgment is not characterizable by rules. If there were rules for judgments, there would have to be rules for the application of judgment, and rules for those rules, *ad infinitum*. Even though judgment does not follow rules it is not arbitrary (Arendt, 1958). Considered judgments in law, medicine, science, and technology are treated as reasonable although they do not follow some formula or recipe.

Nonetheless, a version of rationality, in large part based on science and technology, has won widespread favor in the twentieth century. This is what Weber and others have called **instrumental rationality**. Instrumental rationality is means–end rationality. It involves the search for the most efficient means to reach a given end. Instrumental rationality and the search for efficiency are rightly identified with the technological approach. (Ellul's "technique," with its emphasis on efficiency and the search for efficient means, strongly resembles Weber's instrumental rationality.)

Instrumental rationality also has a close tie with science. August Comte identified the goal of science, not with explanation in terms of essences or natures that the old, metaphysical approach attempted, but with prediction. Predictions are based on causal laws. If a certain thing happens then a certain result will follow. If one strikes a match (provided it is dry) it will light. Instrumental rationality depends on the causal sequences or "if–then" connections of science. If one wishes to reach a certain goal, then one must follow a certain procedure. If one wishes to light a match one must strike it. Means and end mirror cause and effect (Putnam, 1981, p. 175).

One feature of instrumental rationality is that, although it focuses on fitting means to ends, or finding efficient means to reach given goals, it does not evaluate the ends themselves. The choice of ends is itself treated as arbitrary and irrational, or, at least, non-rational. This is in turn tied to the notion that one cannot reason about values and that value judgments are subjective and arbitrary. This viewpoint, popular in our culture, was given classic formulation early in the twentieth century by Max Weber. According to Weber Western culture is being rationalized. More and more areas of traditional thought and action are being structured by instrumental reason. However, the goals or values about which the means are rationally structured are based on irrational decision. There can be no genuine reasoning about values. Here Weber agrees with the existentialists, and sees choice of values as an arbitrary, irrational decision.

Critics of instrumental reason or instrumental rationality disagree with this conclusion. Many of the critics would claim that it is possible to reason about ethics. Classical philosophers such as Plato and Aristotle disagree with Weber and existentialism here.

RATIONALITY, TECHNOLOGICAL RATIONALITY, AND REASON

With a very different approach, the American pragmatist John Dewey (1859–1952) would claim we can reason about values, but his own mode of reasoning is itself means–end reasoning. Ends do justify the means, but not every end is sufficient to justify its means. Ends and means must be fitted to one another.

Against the positivist claim that scientific predictive and possibly causal reasoning is the only form of legitimate reasoning and meaningful discourse, some critics of instrumental rationality would appeal to traditional, meta-physical reasoning. This higher form of reasoning has been characterized in various ways, though these ways are related. Plato claimed that although mathematical reasoning from assumptions or axioms was the training for the philosopher rulers, mathematical reasoning was lower than dialectical reasoning that questioned the fundamental assumptions. Dialectical reasoning examined the forms of values such as justice. (See the discussion of Plato's educational plan for philosopher rulers in chapter 3.)

In the late eighteenth and early nineteenth centuries, Kant, Hegel, and a number of other German philosophers in between contrasted **reason** with **understanding** in various ways. In Kant (1781) understanding is the faculty of both common sense reasoning and scientific reasoning about things and causes. Understanding deals with objects, entities that are delimited in space and time. Objects are finite and have boundaries. They are set against a larger background of space and time, which is supplied by the forms of our perceptual intuition. Things as they are in themselves are not accessible to perception and understanding. However, things as they are for us, things as experienced, are grasped as structured by our perception and understanding. We know *that* the thing in itself exists, but not *what* it is. We only know things as we have organized them perceptually and conceptually. We cannot step outside of our senses or mind to see what things are like when we are not perceiving them, or to think about what things are like independent of our thinking about them.

Reason is the faculty of reasoning concerning notions that are beyond the reach of understanding, such as self, God, and the universe. These latter objects, Ideas of the totality of the universe, and the Ideal of God, are not delimited in space and time, and hence not graspable as objects by reason. They are limits or asymptotes of the sequences of our reasoning, but they are pseudo-objects.

Reason, in Kant's sense here, is the extrapolation to infinity of the notions that understanding applies to finite objects. The immortal soul is infinite relative to time. God is traditionally described as infinite in power,

58

knowledge, goodness, and many other respects. The universe may or may not be infinite in space or in time. Reason in its theoretical form leads to contradiction. Reason, as understanding without the input of empirical experience, is spinning its wheels, so to speak. In this form, when dealing with entities that are not delimited objects or objects of experience, such as God, the soul, and the universe as a whole, reason is the understanding attempting to go beyond its own limits and falling into paradox. Kant calls these contradictions with respect to the universe as a whole the Cosmological Antinomies. Thus, according to Kant, one can refute the notion of the infinite universe, showing contradictions and apparently defending the finite universe. However, one can also refute the notion of a finite universe, showing its contradictions, apparently defending an infinite universe. (The paradoxes of infinite "naive" set theory in mathematics strongly resemble Kant's Antinomies of Reason.)

Similarly, freedom and determinism can both be refuted at a purely theoretical level. However, in a further twist, according to Kant, in the practical, not purely theoretical, realm of ethics, practical reason is able to grasp notions such as freedom that are paradoxical for purely theoretical reason.

Hegel (1770–1831) gave the **dialectic** of theoretical reason a more positive role. The contradictions that reason reaches lead to new formulations that surpass and synthesize the opposing notions that had contradicted one another. For this process, Hegel used a German term (*aufheben*) that means both to abolish and to raise to a new level (sometimes the Latinized term "sublation" is used for it, from the past participle of the verb *tollo*, "to lift"). In contrast to Kant's contradictions of reason, which act as barriers, Hegel's contradictions are the motor that drives reason onward. Although a bit inaccurate, Hegel's dialectic is generally portrayed as starting with a **thesis** (an idea or position), it being countered by an **antithesis** (an opposing, opposite idea), and the two being both absorbed and transcended in a **synthesis** that incorporates the best of each and at the same time raises their fusion to a higher level. Hegel claims that to grasp the limits of reason, as Kant claimed to have done, is in some sense to be able to pass beyond those limits in order to grasp them. This dialectical reasoning shows that reason has no limits of the sort that Kant believed existed. (The slogan of Buzz Lightyear in the Disney cartoon *Toy Story*, "To infinity and beyond," could also be the motto of Hegel and of mathematical theorists of infinite sets, but would be denied by Kant, Aristotle, and mathematicians who demand that all proofs be based on concrete computations.)

59

RATIONALITY, TECHNOLOGICAL RATIONALITY, AND REASON

Note that the dialectic is also no longer an interchange of conversation or a process of thought alone, as in Plato and Kant, but is the very process of reality. Marx, in turn, took over this Hegelian version of dialectic. For Hegel and for Marx society and history are in dialectical process, while for Hegel and Marx's sidekick Friedrich Engels, nature itself is a dialectical process.

The twentieth-century German **critical theorists**, such as Herbert Marcuse (1898–1979) and Jürgen Habermas (1929–), took up the notions of Kant, Hegel, and Marx. They attempted to develop a dialectical approach to the criticism of modern industrial, capitalist, technological society. They saw modern technological society as in the thrall of instrumental reason. Technocratic and positivist notions of superiority of scientific/technological reason and the meaninglessness of traditional metaphysics and ethics are the ideology of modern society. The pushing of questions of ends and values out of the realm of rational investigation and discourse serves to prevent criticism of the implicitly ruling values and the values of the rulers. Marcuse contrasts the traditional metaphysical reasoning of Plato and Aristotle with the limited universe of positivist reasoning and sees the latter as the implicit doctrine of the military industrial bureaucracy. Marcuse sees Weber's sharp distinction between instrumental rationality and consideration of values as an implicit justification of capitalism and bureaucracy. He claims that Max Weber's decisionism, subjectivity with respect to values, and emphasis on social rationalization implicitly serve ultra-conservative ends (Marcuse, 1965). Marcuse even hints that Weber's emphasis on arbitrary decision and charismatic leadership of the ruler points toward fascism, despite Weber's own anti-socialist liberalism. Marcuse even draws parallels between the analytical philosopher's debunking of metaphysical reasoning and the witch-hunting government investigators claiming not to understand the language of their politically radical targets (Marcuse, 1964, p. 192). Marcuse would replace or constrain instrumental rationality with dialectical or philosophical rationality, perhaps even replacing traditional science and technology with a new "liberated" science and technology that serves human values.

Habermas (1987), likewise, sees instrumental rationality as flawed and inadequate as a basis for a good society. Habermas, however, thinks that instrumental rationality is perfectly adequate and appropriate for science and technology. Habermas sees the error not in the application of instrumental rationality to technology but in the extension of instrumental rationality to other areas, such as politics and the family. According to Habermas, scientism and technocracy are the theoretical and political manifestations of

60

this illegitimate extension. Habermas contrasts instrumental rationality as appropriate for the manipulation of things by the individual subject or knower with **communicative rationality**, in which two or more humans interact. Borrowing from the phenomenologist Edmund Husserl, Habermas calls this realm of everyday human interaction the "**lifeworld**" (see chapter 5). What Habermas calls the "colonization of the lifeworld" is the application of technological approaches and instrumental rationality to the realm of human communication. The use of cost/benefit and rational choice approaches to politics replacing the communicative discourse concerning meaning and goals in politics or the replacement of childrearing and education with supposedly scientific behavioral engineering would be examples of this colonization. Habermas's claims as to the dangers of instrumental reason are more modest than those of Marcuse, and concern illegitimate extension and extrapolation of instrumental rationality, not instrumental rationality itself.

Some feminist critics, such as Nancy Fraser, see Habermas's concern for the integrity of authority in the traditional family, immune from intervention by the welfare system and the educational system, as a reactionary position. They see Habermas as in effect defending traditional patriarchy and denying the rights of children against religiously dogmatic or abusive parents (Fraser, 1987). Habermas, on the contrary, himself sees those "essentialist" strands of feminism, which attempt to defend values associated with face-to-face communication, nurturing and childcare, and concern for future generations, as perhaps the most radical contemporary challenge to bureaucratic technocracy.

One problem that both followers of traditional Marxism and devotees of technology studies see in Habermas is his sharp separation of instrumental reason and labor from communication and understanding. Traditional Marxists claim that Marx's concept of social labor is not devoid of human communication (though Marx's account of the role of communication within social labor is hardly fleshed out). Students of technology studies also wish to deny that technological reasoning can be totally separated as instrumental action from the communicative realm of politics or everyday life. Habermas's legitimate concerns about the application of pseudo-scientific and or crudely mechanistic, scientistic social theories to the management and control of social life ("social engineering") are based on a mistaken absolute dualism of labor and communication and of instrumental reason versus communicative understanding. Perhaps it is not surprising that Habermas never analyzes particular examples of technological projects. Indeed, Andrew Feenberg points out that the word "technology" does not occur in the index of the two volumes of his huge *Theory of Communicative Action*. An examination of the interaction of personal values and meanings, political power and persuasion, and the technical, instrumental aspect of technology might undermine his sharp dichotomy (Feenberg, 1995, pp. 78–87).

One source of Habermas's denial of the communicative understanding dimension to technology is his reliance on the logical positivist and Popperian accounts of natural science. Habermas's original accounts of science, technology, and instrumental reason were not cognizant of the post-positivistic treatments of science in American writers such as Thomas Kuhn (1962) and Stephan Toulmin (1961) (see chapter 1). The later Habermas was certainly aware of this work but never did incorporate it into the image of science and technology assumed by his basic schema of instrumental versus communicative action.

It is interesting that Habermas (1970, pp. 50–5) early denied that scientific facts and theories could find a place in the lifeworld. He did this specifically in rejecting the writer Aldous Huxley's appeal for the incorporation of scientific facts and theories into literature (something Huxley had already done in a number of novels). Post-positivistic philosophies of science emphasize the role of paradigms, models, and presuppositions of science. These can function as ideologies and myths in the thought patterns of the lifeworld. Numerous studies of art and science over the past few decades have shown how concepts borrowed from science and technology have been incorporated into imaginative literature and the lifeworld, from non-Euclidean geometry and X-rays in early abstract painting (Henderson, 1983, 1998) to the interest in chaos theory among literary postmodernists (Hayles, 1990, 1991).

Habermas's more moderate position largely replaced Marcuse's utopian but unarticulated call for a new, emancipatory science and technology among practitioners of critical theory. This was in part because Habermas accepted science and technology as they are, and incorporated a number of developments in mid-twentieth-century philosophy and social science. However, it seems reasonable to reject Habermas's strict separation of instrumental action from communicative action and his acceptance of technology as totally value-neutral without having to speculate about a wholly different science that would replace the science we have today. Instead, following the analyses of more recent technology studies, it seems correct to recognize the role of both technical-instrumental reason and political and social values in the social development of technology (Feenberg, 2002, chapter 7). The broader conception of reason, which includes, but goes beyond, the purely formal and algorithmic procedures of mathematics and logic to include contextual judgment, can incorporate reasoning about both the formal-technical aspects of technology and the social and political judgments that are involved in the development of technological systems.

Risk / Benefit Analysis

Risk/benefit analysis is a quantitative means of evaluating technological projects. In structure it strongly resembles Bentham's utilitarianism, in that it sums up positive elements and subtracts negative elements. While Bentham's utilitarianism sums pleasures and pains, risk/benefit analysis sums up benefits and risks. The benefits and risks are generally measured in monetary terms. This is convenient but introduces certain biases into the evaluation. Also, risk/benefit analysis weighs the benefits and particularly the risks by probabilities. A risk is a product of a loss measured in monetary terms multiplied by the probability of the loss occurring. Industrial accidents and the likelihood of individuals getting cancer are examples of the sort of thing that is probabilistically weighted in risk/benefit analysis. In the analysis, one sums up the benefits weighted by the probability of their being accrued and subtracts the risks (the losses times the probability of their being accrued). Risk/benefit approaches are a prime example of technological rationality being applied to the evaluation of technology. The technocratic approach is sympathetic to the apparent rigor and objectivity of risk/benefit analyses (see chapter 3).

There are a number of issues and controversies concerning the applicability and accuracy of risk/benefit analysis. Because of the structure resembling that of Bentham's utilitarianism, some of the problems of the former apply to risk/benefit approaches. The evaluation is solely in terms of consequences. Just as Bentham's utilitarianism does not accept the wrongness of acts or policies on bases other than their consequences (for pleasure in Bentham or for monetary gain and loss in the usual form of risk/benefit), objections by ethicists to **consequentialist** approaches to ethics and of simple utilitarianism apply to risk/benefit analysis.

Some of the problems of risk/benefit analysis are technical and scientific. Estimating the probabilities of accidents is often difficult and speculative. Many complex engineering analyses such as fault trees have been developed. **Fault trees** represent individual failures and their probabilities and are used to calculate the probabilities of sequences of failures in a technological system that might lead to a catastrophic accident, as in nuclear power plant meltdown (Roberts, 1987), and in Failure Mode and Effect Analysis (FMEA), with more focus on manufactured goods (McDermott et al., 1996).

Other problems for risk/benefit analysis are not purely empirical and technical like those above, but more philosophical. Generally the economic benefits of a project are relatively straightforward to assess, but this is not so for the risks. Many forms of harm or risks do not easily or straightforwardly lend themselves to economic evaluation or pricing. One notable example of this is the **value of human life**. Some risk/benefit analyses use estimated future income. This leads to putting a lower value on the death of poor people (with less income) or of older people (with fewer future years of income). One analysis of harm to exposed older asbestos workers downplayed the risk, as these were mostly retired workers who had no future years of job income. Other approaches to valuing life use insurance company actuarial estimates. Here, again, lower income individuals would be likely to buy little or no insurance and thus the value of their life would be very low. Furthermore, some religious and ethical approaches would deny that one could put a comparative monetary value on life at all. If one considers every individual life to be of infinite worth (as some Christian and Kantian approaches would claim), then no monetary benefit of any project, no matter how high, would justify even a low probability of the loss of a single life. (Infinity multiplied by any finite number, even a very small one, is infinity.)

However, defenders of risk/benefit analysis say in reply that we have to make some sort of estimate of the value a technological project, even if it involves, as many do, some small probability of loss of life because of cancer caused by emissions or pollution, injury to workers, or large-scale catastrophic accidents. Thus one must make use of income, insurance, or some other means of valuing life. Advocates of risk/benefit ask how we are to make rational decisions if we reject these sorts of calculations.

Besides the value (or de-valuing?) of human life lost, there are a number of other negative consequences of projects that are difficult to balance against positive monetary benefits. One of these is the aesthetic (or beauty) value of natural scenery lessened by the project. For instance, a power plant may cause air pollution in a national park, lessening some of the scenic views. Nonetheless, some construction firms and government regulators using elaborate models have attempted to assign monetary value to natural scenic beauty.

Another loss difficult to evaluate in monetary terms is the loss of wildlife or of non-commercial species of living things. If one takes the simplest approach, endangered species or living things with no commercial use simply have zero value and their loss counts for nothing. If one takes the most simple and straightforward approach to the commercial value of wildlife, their value may be very low. David Stockman, US budget advisor to President Ronald Reagan, once dismissed the effect of acid rain produced by Midwestern power plants on fish in the Adirondack Mountains of New York by saying that the value of fishing licenses, bait sales, and motel or campground fees for fishermen was very small. This conclusion would seem to be at the opposite end of the scale of valuing nature from the deep ecologists (see chapter 11). Clearly, more complex, indirect ways of estimating the value of wildlife must be used if wildlife and habitat loss is to count for anything in a risk/benefit analysis.

Another area of difficulty or at least complexity for risk/benefit analysis is in considerations of justice. This is also a difficulty for simple Benthamite act utilitarianism. Small benefits to a very large number of people can outweigh huge losses, including that of life (if loss of life is calculated in terms of finite amount of pain) to one or to a small number of people. Manfacturers often calculate the expected amount of money lost in lawsuits to families of injured or dead consumers versus the cost of more extensive testing of the product or modification of the product. The Ford Pinto case is a classic example of this, in which the manufacturer did a cost/benefit analysis of manufacturing new gas tanks, less likely to explode on impact, versus the cost of lawsuits concerning injured or dead Pinto riders. Drug companies traditionally calculated the minimum of the sum of the curves of the cost of more frequent testing of samples of the product, and the cost of lawsuits from injuries or illness caused by defective instances of the product.

Often the recipients of the benefits and the sufferers of the losses involved in the risks are different groups of people. The investor recipients of economic profits or the consumer benefits of electric power or manufactured goods from a power plant or a factory often live far from the location of the plant or factory, while the sufferers from pollution, radiation, or other risks live near the power plant. Critics of multiple, mercury preservative laden vaccinations for young children object to the alleged cost in autism for a few children, despite the benefits of protection from disease for many. A simple summation of total risks and benefits ignores the problem of distributive justice. Some analysts have supplemented risk/benefit analysis with distribution considerations.

Some opponents of risk/benefit analysis note that the practitioners and advocates of risk/benefit approaches generally use them to justify the technological projects that are analyzed. Risk/benefit advocates are often claimed to be overwhelmingly also advocates of reduced government regulation of the enterprises examined. Furthermore, the rhetoric of "risk taking" is used in advertisements claiming that "rugged individualist" American pioneers were risk takers, while modern consumers are cowardly in their risk avoidance. Of course, this rhetoric falsely equates voluntarily and knowingly accepted risks of travel by pioneers into unknown territory with the involuntary and often unknowingly suffered risks of pollution, radiation, or poorly made and defective products. That corporate advertisers have sometimes made use of this misleading rhetorical praise of risk does not, of course, in itself say that technical risk/benefit analysis is mistaken.

Often it is argued that the risk of the pollution or radiation being examined is less than some more mundane and commonly accepted activity that has a generally unrecognized risk (such as exposure to radiation from frequent high altitude airplane travel or household radon exposure in some regions). Furthermore, it is noted that the proponents of risk/benefit often castigate the public as "irrational" in their fear of, say, nuclear power, and their acceptance of other low-level radiation sources. Risk/benefit proponents also use research on the inaccurate probability evaluations that people in general make in everyday situations (Kahneman and Tversky, 1973). Implicit in much of this criticism of the irrationality of the general public is the suggestion that only the scientists, engineers and risk/benefit analysts are qualified to make reasonable judgments about acceptability of technological projects (Perrow, 1984, pp. 307–15). That is, the contrast of the ignorant and irrational public with the cool and rational risk/benefit analysts implicitly supports technocracy (see chapter 3). However, Kahneman and Tversky find that, in their informal assessments of probability in real life situations, even experts in probability theory commit the same sort of fallacies as do ordinary people.

In fact, many proponents of risk/benefit have themselves accepted research on or have themselves investigated the psychological dimensions of risk and found patterns of evaluation behind the supposedly "irrational" judgments of the public that justify some of those judgments. For instance, involuntary risks are considered less tolerable than voluntary risks. Unknown or unfamiliar risks are less tolerable than known or familiar ones. Risks with catastrophic potential (potential for a major disaster) are considered less acceptable than risks that cause scattered harm or loss of life spread widely over time and space. Finally, risks that unfairly or inequitably distribute the risks and benefits (to different groups or putting the risk on future generations) are less acceptable than risks that are equally distributed over the population (Lowrance, 1976, pp. 86–94; Slovik et al., 1981). Traditional pure risk/benefit analysis would consider all these considerations to be irrelevant to risk as such.

Some analysts contrast "perceived risk" with "real risk," and some say that popular perceptions of risk, even if unscientific, have to be taken into account politically. However, it is not self-evident that consideration of the inequity, catastrophe potential, or involuntary nature of a risk is irrelevant to the "real risk" measured, say, purely in terms of decrease of average life-span. Some would distinguish between calculation of risk as such and the judgment as to the "acceptability" of the risk. The latter can reasonably consider issues such as the voluntary nature or inequity of the risk.

Regardless of one's evaluation of the degree of "rationality" of taking into consideration the above factors in considering the acceptability of risk, one can argue that policy decisions concerning societal risks are necessarily political. In policy decisions we are dealing with societal decision-making, not the psychology of single individuals. Community consensus building is necessary for social decisions. This process itself necessarily has a political element. The group decision-making devices are its politics (Rescher, 1983, pp. 152–6).

Social constructionists (see chapter 12) claim that all risk evaluation is socially constructed. It is claimed that power relations, negotiations, and political ideologies permeate the entire risk/benefit analysis. Considerations such as those noted above, concerning the apologetic uses of comparisons of everyday risks with risks of high technology projects and concerning the technocratic tendency toward denigration of the rationality of popular estimates of risk, support the social constructionist position. However, the social constructivist goes further and claims that the entirety of the methods and data involved in risk analyses is socially constructed and pervaded with political bias. The leading anthropologist Mary Douglas, in concert with the neo-conservative policy analyst Adam Wildavsky, goes as far as to claim that fears about air pollution are simply primitive taboos concerning pollution and purity, and have no relation to physical reality (Douglas and Wildavsky, 1982).

One way of attempting to disentangle the scientific aspects of risk/benefit calculations from political or social judgments concerning opposing or supporting technological artifacts or projects based on these analyses is to distinguish "risk/benefit analysis" from "risk management." This would sharply distinguish the "scientific" from the "political" aspects of risk evaluation. The relationship of social views or political biases to risk/benefit analysis is more complicated than this, however.

It is true that not just in policy decisions but also within the risk/benefit analysis itself there are areas into which one's social biases can enter. Judgments must be made concerning which low probabilities to discount as "effectively zero," which threshold levels of statistical evaluation to use, which models of extrapolation to humans from experiments on animals to use, as well as many others. One's biases, whether to play down risks or accentuate them in any particular case study, can affect decisions made concerning risk calculations. Thus one cannot isolate politics within the policy decision aspects of risk management and keep the science pure and unaffected by social attitudes. However, this is not to say that there is no place for scientific objectivity in risk analysis. As we have seen in our survey of the philosophy of science, a mechanical or algorithmic approach to induction or science is not viable (see chapter 1). We find that a purely mechanical method of risk analysis is impossible and that, just because social attitudes can affect scientific judgments at crucial points, this does not mean that the risk analysis is totally arbitrary or the complete hostage of social biases. Once one realizes where judgments concerning thresholds and extrapolations need to be made, one can examine what biases may have entered the calculations and criticize them. Thus, while the totally unbiased and mechanical evaluation of "real risks" is unrealistic, the social constructionist exaggerates the extent to which risk evaluation must be simply an expression of bias and prejudice (Mayo and Hollander, 1991).

Conclusion

We have seen a number of different sorts of reason. The formalistic version identifies reason with deductive logic. Euclid's geometry was the model both for Plato and for the seventeenth-century rationalists. Some later identified reason with a formal inductive logic, which in Carnap approaches, in its *a priori* structure, a deductive system. Others identify reason with instrumental or technological reasoning, the adaptation of means to end. Kant and Hegel, in different ways, contrasted ordinary logic with transcendental logic, and reason with the more lowly understanding. In its romantic extremes, this transcendental reason can become a quasi-divine intuition or the aesthetic judgment of an artist. Economic, calculative reason is the reason appealed to by risk/benefit analysts. Formalistic, economic, and instrumental reason all underestimate the need for non-rule-bound judgments.

Critical theorists contrast transcendental and dialectical reason with formalistic, quantitative, and technological, manipulative, instrumental reason.

Study questions

1 Does formal (mathematical–technological) rationality completely capture what it means to be rational?
2 Is there "metaphysical" or "dialectical" rationality above and beyond instrumental or technical rationality?
3 Is risk/benefit analysis acceptable as a means of evaluating technological projects, or is it to be rejected because of its neglect of rights and justice except in terms of monetary consequences?
4 Is the valuing of life (as well as non-human living things and natural scenery) in monetary terms to be rejected, or must we use it because it is the only method for balancing lives (and species or scenery) lost against the benefits of technological projects?

5

Phenomenology, Hermeneutics, and Technology

The logical-linguistic **analytical** approach and the **phenomenological** approach are two major trends in twentieth-century philosophy. Originally the logical analytic mode originated in and dominated English-speaking philosophy (although the Austrian logical positivists, Berlin logical empiricists, and Polish logicians were a central part of this trend), while the phenomenological approach dominated "**continental**" philosophy (particularly French and German). The very term "continental" (for the mainland of Europe, in contrast to Britain and Ireland) shows that this way of putting the dichotomy is due to British philosophers and their American followers. In recent decades there has been a great deal of "bridging" of the Anglo-American/continental split, with analytical and pragmatic philosophy growing in Germany while a substantial minority of American and British philosophers are making use of French and German philosophy. Furthermore, some versions of French structuralism and postmodernism have conceptual resemblances to Anglo-American linguistic philosophy, even if styles, rhetoric, and attitudes are so different as to make communication difficult.

Phenomenology is the description of experience. Phenomenology attempts to describe pure experience, avoiding the common tendency to attribute features to our experience that we "know" (or think we know) are in the object. (This tendency is what William James called "the psychologists' fallacy.") We might compare the phenomenologist to the impressionist painter, who paints objects as perceived in a particular light, time of day, or shadow, not the objects with the colors that they are "supposed to have" in standard conditions of bright sunlight.

The original phenomenology was that of the German Edmund Husserl (1859–1938). He was trained in both mathematics and psychology, and thus was more sympathetic to, and knowledgeable about, science than most of his immediate disciples. Husserl claimed to describe experience without theory or presuppositions affecting the description. For Husserl and other phenomenologists all consciousness is consciousness-of-something. This feature, the directness of consciousness, is called **intentionality** (the phenomenologists' emphasis on intentionality puts them in strong opposition to both logical positivists in philosophy and behaviorists in psychology). The question of the existence of the objects of experience was "put in brackets" or suspended by a so-called "phenomenological reduction." The phenomenological description of experience includes both perceptual experience and the conceptual aspects of experience. Husserl and his close followers claimed to extract "essences" of experience through what Husserl calls "eidetic intuition." (*Eidos* is Greek for Platonic form. See the discussion of essences in chapter 2 and of Plato in chapters 2 and 3.)

Husserl's most influential student, and soon his competitor for intellectual dominance of German (and later French) philosophy, was Martin Heidegger (1895–1976). Heidegger's version of phenomenology emphasized lived existence and claimed to avoid the Platonistic formalism of Husserl, as well as the spectator-like approach to knowledge that Husserl shared with Descartes, the British empiricists, and much of traditional Western philosophy. Heidegger rejected the ostensively detached and neutral standpoint of Husserl's phenomenological reduction. Furthermore, Heidegger replaced Husserl's rather Platonic–Aristotelian notion of essences and eidetic intuition with an interpretive involvement in the world. Heidegger replaced the abstract categories of former philosophy with "existentials," and used something akin to moods rather than formal abstractions as the fundamental orienting structures of human experience.

Husserl developed and modified or supplemented his phenomenology in his later work. In his *The Crisis of European Science and Transcendental Phenomenology* (1936), Husserl discussed Galileo and the founding of modern physical science. Husserl added the notion of the "lifeworld," the world of ordinary lived experience that is in the background of scientific abstraction. Some claim that this work was itself an attempt to respond to criticisms and developments of phenomenology by his student Martin Heidegger. Heidegger rejected Husserl's presuppositionless phenomenology.

In the 1960s American followers of the French philosophers influenced by Heidegger's phenomenology (Jean-Paul Sartre and Maurice Merleau-Ponty in their writings in the period of the 1940s and 1950s) dubbed the new movement "**existential phenomenology**." Although this label misses some of the subtleties and complexities of Heidegger's thought, it does give a rough characterization of the dominant new approach to phenomenology. Existentialism, to use Sartre's slogan, takes existence as prior to essence. It emphasizes the concrete, unique individual rather than general forms or natures. In this it resembles nominalism and empiricism despite the great difference in style and topics. Existentialism focuses on personal life, while empircism mainly focused on empirical science. But all these trends reject Plato's forms and extreme rationalism. Existential phenomenology is more concrete than the original form of Husserl's phenomenology. Heidegger's own work is more properly called "**hermeneutic phenomenology**," because it incorporates insights from the theory of interpretation of texts and culture (see more on hermeneutics below).

Husserl himself in *Crisis* applied phenomenology to the sciences. In his earliest work he had applied phenomenology to logic and arithmetic. A number of Husserl's immediate followers applied phenomenology to mathematical physics, to psychology, and to medical diagnosis. Husserl emphasized that the abstractions of science are an **idealization** of the concepts of lived experience (the lifeworld). Points and perfectly straight lines in mathematical physics are idealizations produced by successive approximation from the ordinary volumes and shapes of experience. The British-American process philosopher Alfred North Whitehead (1861–1947) in his "method of extensive abstraction" developed around 1920, and even the mid-nineteenth-century British empiricist John Stuart Mill (1806–73) in a little-noticed passage in his *A System of Logic* (1843), developed similar methods of idealization (see the discussion of process philosophy in the section on actor-network theory in chapter 12). For Husserl, the claim that the idealizations of mathematical physics are real and that ordinary experience is somehow illusory completely inverts the relationship between the life world and science. The fundamental starting point is **lived experience**. The abstractions of science are constructs that allow us to predict and control but are not, somehow, more real than the objects of ordinary experience (Whitehead called this "the fallacy of misplaced concreteness"). Both the objects of ordinary experience and the objects of science are objects of experience that can be described phenomenologically.

Hermeneutics means "interpretation." Hermeneutics started in the interpretation of Biblical texts. German theologian and philosopher Friedrich Schleiermacher in the early nineteenth century developed hermeneutics into the discipline of interpretation of texts in general. German philosopher Wilhelm Dilthey in the late nineteenth century expanded hermeneutics to encompass understanding of human behavior and culture. Heidegger incorporated notions from hermeneutics into his own version of phenomenology (hermeneutic phenomenology). Hans-Georg Gadamer greatly developed the hermeneutic side of Heidegger's endeavor. Hermeneutics, in sharp contrast to the early version of Husserl's phenomenology, does not claim to approach matters from a presuppositionless standpoint. Instead, hermeneutics claims that we approach a text with "prejudices" (to use Gadamer's term) or pre-understandings. These enable us to interpret the text, and in turn allow us to examine those preliminary orientations. This is the so-called "**hermeneutic circle**." Although one depends on one's initial understandings to interpret the text, the subsequent interpretation helps us to readjust those understandings.

Hermeneutics has only in the past few decades been applied to the natural sciences. Patrick Heelan (1983) and Don Ihde (1998) are two American philosophers who have engaged in this task. Previously hermeneutics of science had meant cultural interpretation of science, but Ihde has developed a hermeneutics *in* science; that is, an account of the interpretive activity of scientists in instrument reading.

Don Ihde's Technology and Lifeworld and Expanding Hermeneutics

The most lucid and easily understandable application of phenomenology to technology is to be found in the works of the contemporary American philosopher of technology, Don Ihde. In his *Technology and the Lifeworld* (1990) and other works, Ihde has concentrated on the role of scientific instruments in observation. Phenomenology of perception is the starting point of the analysis, but Ihde emphasizes that technological devices mediate scientific perception. Ihde embraces and incorporates the work of the instrumental realists (see chapter 1). Like the instrumental realists, he does *not* wish to claim that those objects of science that are not directly observed in everyday perception are somehow "theoretical entities" that lack reality or have an abstract reality different from that of ordinary objects. Instead, unlike many continental philosophers, Ihde claims that the objects of the most technical science are objects of perception (albeit a technologically extended perception) just like common sense objects.

Ihde notes that there are two different modes of instrumental perception. In one the observer and instrument are united in contrast to the object of science. He illustrates it by:

(I – instrument) – object; as in (I – telephone) – you

In the other the observer is reading the instrument–object complex. The former describes the phenomenology of instruments as being incorporated into our embodiment. The latter describes the situation in hermeneutics. Ihde illustrates this situation as:

I – (instrument – object); as in engineer – (instruments – atomic pile)

In the instrument reading situation, the object is "read" through the instrument. It is not literally "seen" as through a microscope or telescope; instead, the instrumental readout allows us to interpret the object.

Recall that traditional hermeneutics started with the interpretation of sacred texts and expanded to the interpretation of texts in general (Schleiermacher) and then to culture and history in general (Dilthey). However, even in its later generalization (which held from the late nineteenth century until the late twentieth century), hermeneutics was (and still mainly is) conceived of as a purely humanistic discipline. Dilthey held to the distinction between the human sciences (*Geisteswissenschaften*) and the natural sciences (*Naturwissenschaften*). The two kinds of science were contrasted in terms of (interpretive) "understanding" in the human sciences and (causal or formal) "explanation" in the natural sciences. Other late nineteenth-century figures, the Southwest German school of followers of Immanuel Kant (Windelband and Rickert) contrasted the sciences of the individual (idiographic) with the sciences of the lawful (nomothetic) and identified this with the split between the humanities and natural sciences. The idiographic and hermeneutic approaches, focusing on the unique individual and its interpretation, were appropriate to history and the humanities, but the natural sciences involve deduction of descriptions from universal laws.

One of the developments of late twentieth-century thought was to break down this sharp dichotomy. With the rise of post-positivistic philosophy of science, most famously with Thomas Kuhn, but also with Michael Polanyi, Norwood Hanson, Stephen Toulmin, and a host of others since the late 1950s and 1960s, the image of science as a purely formal, deductive machine, cranking out predictions from universal, formal laws, was replaced by images of science involving informal paradigms and tacit presuppositions, models and context-dependent inferences (see chapter 1). Thus the "nomothetic" image, accepted as the accurate description of natural science by the hermeneutic and neo-Kantian traditions as well as by the logical positivists,

Box 5.1

Heidegger and technology

Martin Heidegger (1889–1976) was one of the most influential philosophers of the twentieth century, probably the most influential philosopher from continental Europe. Heidegger, particularly in his later writings, had a central concern with technology. Students of Heidegger such as Hannah Arendt (1906–75), Hans Jonas (1903–93), and the Marxist Herbert Marcuse (1898–1979) were among the other major philosophers of technology of the mid-twentieth century.

Early in his first major writing, *Being and Time* (1927), Heidegger analyzes our comportment to objects in terms of two modes of being, ready-to-hand and present-to-hand. The ready-to-hand mode is that of tool use. The object exists in its role in our action. This mode resembles the approach of pragmatism in American philosophy. This contrasts to the present-to-hand mode, which is the usual notion of objects as substantial entities observed or thought of as independent and over and against us. Traditional philosophy treated all objects as present-to-hand. Objects are perceived by the senses (by empiricists) or conceived by the intellect (by rationalists) as separate and distinct from us. In contrast the ready-to-hand object of use is not perceived as an independent entity but is instead the means through which we work and act. The ready-to-hand object such as a tool is transparent to us when we use it properly. Only when it fails to operate properly or breaks do we become aware of it as present-to-hand. Heidegger famously uses the example of the hammer. When we are using it to drive nails, our focus is on the successful driving of nails into wood. Only when the hammer breaks do we focus our attention on the hammer itself, rather than the result for which it is normally used.

The tool-like mode of existence was central to Heidegger's earlier philosophy. In his later philosophy technology itself becomes a subject of reflection. Heidegger claims that modern technology defines the present epoch of humanity just as religion defined the orientation to the world of the Middle Ages. Modern technology differs from previous crafts (although it grows out of them) insofar as it "enframes" or "stamps" everything with its orientation. All of nature becomes a "standing reserve," a source of resources, in particular a source of energy. This enframing cuts us off from appreciating non-technological ways of apprehending the world and also obscures the character of the all-encompassing technological enframing itself. That is, we become so entangled in the technological way of thinking or the technological attitude that alternative ways of thinking, those of archaic, pre-industrial peoples, become inaccessible to us. Furthermore, we become so immersed in technological approaches to the world that we are not even aware that the technological attitude is one approach of many. It is assumed to be identical with sound or correct thinking.

For Heidegger, humans are not in control of technology. Instead, technology is the destiny of humans in our age. Heidegger describes our present relation to technology in the same way proponents of autonomous technology describe it (see chapter 7). Technology is not in human control. For people to demand to "get in control" of technology simply reinforces the technological attitude. For technology's approach to nature is to control it. Technology is so definitive of our age as to preclude any attempt to retreat to a pre-technological society or culture.

Heidegger does think that it is possible to achieve a "free relation to technology." Exactly what this is is disputed. He thinks it is possible to gain an understanding of the essence of technology, something not achieved by traditional philosophy or Christian religion. Once one grasps the essence of technology it is possible to use technology without being caught up in it.

In much of his work Heidegger contrasts traditional crafts and the peasant way of life with work and life in modern technological society, much to the detriment of the latter. He uses examples of a Greek temple, a silver chalice, and a traditional wooden bridge, contrasting them with a modern power plant or superhighway. Heidegger's preference for and praise of rural, peasant life and dislike of cities suggests that he is an anti-technological romantic. It seems that in rural and non-technological settings we grasp the genuine meaning of things. But this is misleading, given that he claims that technology characterizes our age and we cannot return to pre-technological ways. In some passages Heidegger claims that technological artifacts themselves can be occasions for us to grasp being. Heidegger uses a pitcher and an old bridge as examples of the nexus of unification of earth and sky, humans and gods, in their making and their use. However, at one point, contrary to his usual use of archaic and rural examples, he claims that the modern highway interchange can also function in this way as a focus of aspects of being.

has, for many philosophers of science, been replaced by another image of science more amenable to and similar to hermeneutic interpretation.

Hermeneutic interpretation of the history of science and science as a cultural phenomenon goes beyond the work of the early hermeneutic thinkers but is not inconsistent with their approach. However, a more radical expansion of hermeneutics that Ihde defends is a hermeneutics of the *objects* of science, not simply the culture, history, and psychology of scientists. That is, science itself can be seen as having an interpretive dimension to which the early tradition of cultural hermeneutics was blind.

As suggested by the second schema of Ihde above, the "reading" of instruments, including the implicit interpretation of those readings within a theoretical framework, is a form of hermeneutics of the objects of science.

Dreyfus and the Critique of Artificial Intelligence

Hubert Dreyfus, an American follower of Heidegger, has given what is perhaps the most influential application of the phenomenological approach to technology in his critique of artificial intelligence. Despite initially highly negative reactions from some members of the artificial intelligence research community, including attempts to prevent publication of some of his early writings on this topic, Dreyfus's work, unlike much philosophy of technology, has had a direct influence on a number of computer scientist practitioners and has led to modifications of their strategies.

In his early RAND Corporation report *Alchemy and Artificial Intelligence* (1965) and in his *What Computers Can't Do* (1972), later revised as *What Computers Still Can't Do* (1992), Dreyfus argued that classical **AI** (**artificial intelligence**) was based on mistaken assumptions about thinking and meaning that were shared by early modern philosophers such as Descartes and the British empiricists. At the time of Dreyfus's first writings, the "**classical**" or **symbolic processing** approach to AI was the only one. Since then, with the popularity of neural network theory, symbolic processing AI has been called "classical." Nowadays it is sometimes called GOFAI or "good old fashioned artificial intelligence." In the classical approach, thinking is claimed to deal with combinations of discrete bits, and reasoning to consist of the manipulation of symbols according to explicit rules. Dreyfus claimed that although the symbolic processing approach is appropriate for formal logic and mathematics, for areas such as understanding natural language, classical AI does not appropriately model perception and everyday reasoning.

Thinking and judging without explicit rules

Dreyfus argues, following Heidegger, but also the later philosophy of Ludwig Wittgenstein (a source of most ordinary language philosophy in Britain and the USA), that the rules at the basis of thinking cannot be fully formalized or made fully explicit. Behind our capacity to judge and reason are tacit or pre-reflective orientations. Traditional rationalism, deriving from Plato, assumes that all reasoning can be made explicit and mathematically formalized (see chapter 4). A few philosophers, a definite minority, dissented from this tradition. Aristotle, considering ethical judgment, said that ethics cannot have the exactitude of mathematics, and that judgment is necessary in practical wisdom. Blaise Pascal (1623–62), a mathematician turned religious devotee and enemy of Descartes, famously said, "The heart has reasons that reason does not know." Pascal contrasted the "geometric spirit" with the "spirit of finesse" that uses non-formal intuitions of appropriateness. The early twentieth-century French physicist-philosopher-historian Pierre Duhem agreed with Pascal and claimed the inductive and deductive methods of science were inadequate to account for decisions as to whether to reject or to slightly modify their theories in the face of counter-evidence (see chapter 1). However, Pascal and Duhem were outside the mainstream and largely ignored on this issue until the past few decades (see chapter 4 on the rehabilitation of the role of judgment in Aristotle and Kant in recent accounts of rationality).

Dreyfus claims that human rationality involves the ability to apply rules to particular contexts in a manner that cannot be fully formalized. If one assumes rules to apply the rules, one is driven to assume further rules to apply these rules and into an infinite regress. What the Hungarian-British chemist, social theorist, and philosopher Michael Polanyi (1958) called "tacit knowing" or what the American pragmatist John Dewey (1931) called "the context of thought" must be involved in human judgment.

Holism, the field of consciousness, gestalt, and horizons

Husserl and the phenomenologists follow the so-called **gestalt psychologists** in claiming that perceptual experience is structured in terms of figure and ground relationships (a gestalt is an organized figure or shape). The ground is the background within which we experience a particular delimited object, shape, or figure. This ground, which is often out of conscious attention, or "tacit" (in Michael Polanyi's terms), structures our perception of the figure that occupies our "focal" awareness. Our perception and cognition has for Husserl **"horizons."** There is the "outer horizon" of the indefinite distance in perception or the indefinite iteration of some operation in cognition. There is also the "inner horizon" of the un-focused-upon aspects of our focal objects. Consciousness is not simply a display of equally explicit features and facts, but has a **holistic** structure (see the discussion of holism in chapters 1 and 11).

The **field of consciousness** with its outer and inner horizons and its internal structure and organization is not simply an assembly of bits or of atomic sense data. Instead it is a holistic entity, in which the background structures and situates figures that are the center of attention, and receding horizons of possibly indefinite extension of perception or unending reiteration of rules of conception are part of experience itself. In this way consciousness as described by phenomenology differs from consciousness or perception as described by traditional AI, or the British empiricist tradition for that matter. Even the Austrian logical positivist Rudolf Carnap, in his earliest major work, *The Logical Construction of the World* (1928), based his constructions on gestalt perceptions rather than the impressions of Hume, elements of Ernst Mach (1838–1916), or sense data of Bertrand Russell. However, the Anglo-American development of positivism dropped this aspect of the early Carnap.

Embodied thought and the lived body

According to Dreyfus, thought and intention assume a body, not the mechanical body, as described by physics and chemistry, but a "**lived body**," in the sense of Maurice Merleau Ponty (1907–61), a follower and developer of the ideas of the later Husserl. The "lived body" is a notion different from that of the pure mind as a spiritual or mental entity, and also different from the body considered as a mechanical body of physics. The lived body is experienced "from within" as an intrinsic part of our orientation to the world and objects. It contrasts with the body studied by biology and medicine "from without" as an external object confronting us. Descartes, who in the seventeenth century formulated mind/body dualism, considered the body a mechanical apparatus and the mind a purely spiritual substance or entity different in nature from the body. The lived body falls outside of either member of Descartes's dichotomy. Dreyfus claims that the intellectualism of traditional AI and some of the philosophy associated with it considers thought something that can be done by a disembodied thinking apparatus. In Dreyfus's account, perceiving organized wholes is a bodily practice and skill. The relations of the temporal rhythms of the body to the recognition of melodies and the coordination of the body in relation to its surroundings involved in eye movements in seeing are examples of this.

The upshot is that we orient ourselves to objects or experiences through our embodiment. Our bodily dispositions, with their propensity for action, organize our experience. Similarly, our emotions, attitudes, and goals structure our experience. The problem with classical AI implicitly assumes that perception and thinking are simply arrays of discrete qualities, attributes or features of objects, and facts, manipulated or processed according to explicit rules.

Neural networks or connectionism to the rescue?

Many in the AI community saw the revival of **neural network theory** or **connectionism** in the 1980s as a solution to the problems into which classical or symbolic processing AI had run.

The use of neural nets or connectionism to model perception, as exemplified by the perceptrons of Arturo Rosenblatt, was quite an active field in the early 1950s and 1960s. An alleged proof of the impossibility of neural nets working beyond a certain level of complexity by Marvin Minsky and Seymour Papert (1969) dampened interest, until the problems of symbolic processing AI led to a revival of connectionism.

Connectionism does not involve modeling thought as representation. Neural networks, as their name suggests, were inspired by modeling to some extent the neural systems of the brain, rather than modeling the rationalist-deductive model of thought. Devices called **perceptrons**, attempting to model the function of the retina and brain, and having a set of connections from the electronic sensors to the recording and storage device, randomly form connections that are reinforced by success in training. Dreyfus is sympathetic to this aspect of neural nets. Dreyfus is quite favorable toward neural network modeling, although insofar as it involves conceptualization of thinking as processing of information he is critical of it.

Neural networks and connectionism have been quite successful with perceptual recognition, but have been much less so with deductive reasoning, the strong point of symbolic processing AI (which, however, has been quite weak on perceptual recognition and related tasks). Dreyfus notes that the training that neural networks undergo is often highly focused by the selection of relevant attributes by the human trainer. Sometimes when not having a pre-selected set of attributes to train upon, the neural nets "learn" to

Box 5.2

The tangled roots of neural network theory: philosophical and psychological

The basic idea of neural networks has been around since the late 1940s. The neurophysiological model for the early version of neural networks was suggested by the work of psychologist Donald O. Hebb (1949). Hebb described "reverberating circuits" in the brain as the basis of explanation of thought. There was a less well known philosophical generalization of Hebb's work by the otherwise famous, Austrian, free market, conservative economist-philosopher Friedrich Hayek (1952). Hayek attempted to eliminate phenomenal sensory qualities by reducing them to a system of relations, rather along the lines of Mach and Carnap (see chapter 1). Hayek conceived of the mind as without a central organizing principle, but made up of competing neural elements, rather the way Adam Smith's (1776) economic "invisible hand" of the market emerges out of individual competition.

Warren S. McCulloch and Walter H. Pitts developed the logical theory of neural networks. Interestingly for philosophers, and perhaps surprisingly for cognitive scientists, McCulloch was a follower of the medieval scholastic philosopher John Duns Scotus (McCulloch, 1961, pp. 5–7). The realist theory of universals of Duns Scotus was followed by the American pragmatist philosopher Charles S. Peirce. (In a little-noticed turn of twentieth-century philosophy, Martin Heidegger (1916), the German hermeneutic phenomenologist, and Peirce (1869), the pioneer of symbolic logic, were both early influenced by a work on "speculative grammar" purporting to be by Duns Scotus (Bursill-Hall, 1972). The scholasticism of Duns Scotus's seventeenth-century followers was so rejected and ridiculed by early modern philosophers such as Bacon that "Duns" was the origin of our term "dunce," for fool, yet his views lie behind two major branches of philosophy of language in the twentieth century.) Walter Pitts, McCulloch's mathematical amanuensis, was living as a 13-year-old runaway in Chicago when he ran into a library to hide from bullies. Next to his hiding place was Bertrand Russell's three-volume *Principia Mathematica*, deriving mathematics from logic, which Pitts read in a few days. He wrote to Russell in England with criticisms and corrections. According to one account Pitts happened later to meet Russell in a Chicago park (where he initially took the then somewhat seedy looking Russell for a fellow street person), and attended a lecture of Russell's in Chicago. Russell recommended Pitts to the Vienna Circle logical positivist Rudolf Carnap at the University of Chicago. Pitts then went to MIT and worked with Norbert Wiener (1894–1964), himself an ex-prodigy (Heims, 1980), on cybernetics feedback mathematics, as well as with Jerome Lettvin, who wrote, among many articles with Pitts and McCulloch, a brilliant analysis of the visual system of the frog (Lettvin et al., 1959).

Pitts, looking for an absent father figure, was everyone's son at MIT and did the detailed mathematical working out of everyone else's ideas. Later, Wiener's wife accused Pitts and Lettvin of sleeping with her daughter (a lie), as she was jealous of Wiener's connection with these fellows. She also disliked McCulloch's drinking and extramarital sex life. The group broke up at the moment of its greatest promise, without Pitts or McCulloch ever learning the reason for Wiener's rejection of them (Conway and Siegelman, 2005, chapter 11). Partly out of despair at Wiener's unexplained rejection, Pitts later drank and drugged himself to death in a flophouse in Cambridge (Heims, 1991, pp. 154–5). This tragic death and loss of a treasured colleague led Lettvin to engage in public debates with Timothy Leary, the devotee of LSD, wherein Lettvin opposed encouragement of personal experiment with LSD.

solve a recognition problem by using what turns out to be a trivial and irrelevant attribute that happens to work in the particular examples on which they are being trained.

"Heideggerian AI"

Dreyfus's critique of AI research has had genuine impact on some programs and investigators in the AI field. Despite the initially highly negative reaction at MIT and Pittsburgh to Dreyfus's original debunking of failed AI predictions, some researchers have in recent decades engaged in "Heideggerian AI." Terry Winograd and Fernando Flores (1986) have advocated an approach to AI taking into account Heideggerian phenomenology and hermeneutics. Philip Agre and David Chapman (Agre and Chapman, 1991; Agre, 1997) in their interactive programming have also incorporated Heideggerian insights into everyday coping and orientation toward the world in their research.

Conclusion

These examples, the work of Don Ihde and of Herbert Dreyfus in particular, show that the phenomenological and hermeneutic approach to technology has much to contribute to the description and understanding of technological practice, and even to the improvement of technology itself.

Study questions

1 Do you think that description of our perceptions and intuitions is an effective way to understand technology? Is it sufficient? If not, what more is needed? Do you think that logical proof and argument is a better or worse approach? Should the two supplement each other?

2 Do you think that scientific instruments are extensions of human bodily perception? Are instrumental observations incorporated into our bodily experience?

3 When we read a meter or dial in a scientific instrument, are we in effect "reading nature," or is our dial-reading and intellectual act separated off from nature?

4 Do you think that digital computers can perceive and be conscious? Does a computer need a body in order to be aware?

5 Is Dreyfus's denial of the prospects of success of the project of artificial intelligence refuted by the development of neural network theory?

6 Do you think that we can achieve "a free relation to technology" as Heidegger asserts? What would this be like?

6

Technological Determinism

Technological determinism is the claim that technology causes or determines the structure of the rest of society and culture. Autonomous technology is the claim that technology is not in human control, that it develops with a logic of its own. The two theses are related. Autonomous technology generally presupposes technological determinism. If technology determines the rest of culture, then culture and society cannot affect the direction of technology. Technological determinism does not, on the face of it, presuppose autonomous technology. It could be that free, creative inventors devise technology, but that this technology determines the rest of society and culture. This would leave the inventors outside of the deterministic system as free agents. However, if science has a logic of its own, as Heilbroner claims in defending technological determinism, and technology is applied science, then the inventors are not free to develop technology as they see fit, and we are back to autonomous technology.

According to technological determinism, as technology develops and changes, the institutions in the rest of society change, as does the art and religion of the society. For instance, the computer has changed the nature of jobs and work. The telephone led to the decline of letter writing, but the Internet has changed the nature of interpersonal communication again, leaving written records unlike the telephone. The automobile affected the distributing of the population by leading people to move from the city to the suburbs and leaving central cities impoverished. The auto in the 1920s and 1930s even changed sexual habits of youth by making available the privacy of the auto outside the house. A juvenile court judge called the auto a "house of prostitution on wheels," and FBI Director J. Edgar Hoover called early tourist courts (motels) "little more than camouflaged brothels" (Jeansonne, 1974, p. 19).

A set of technological determinist claims about earlier history involves the rise and fall of feudalism; that is, the social system of the Middle Ages. The introduction of the stirrup from central Asia to Europe made possible the mounted armored knight. Without the stirrup the charging warrior was knocked from his horse by the impact of his lance hitting an opponent. The support of the knight (devoting his time to military practice), the horse, the more and more elaborate armor for both, and the knight's retainers was expensive. The feudal system of peasants providing a part of their agricultural produce in exchange for protection supported the expense of the horse and armor of the knights (White, 1962). The feudal knights further developed the ethic of chivalry and courtly love. The whole system, military, economic, and cultural, is claimed by to have collapsed because of the later introduction of another Asian technology, gunpowder from China. The armored knight fell to the shots of guns and the castle became prey to bombardment by the cannon. (Needless to say, various historians have criticized the details of this neat technological determinist account.)

I think we shall see that one should apply technological determinism and cultural determinism on a case-by-case basis. In some situations the technical and physical aspects of the technology propagate major changes in culture. In other situations the cultural and value orientations of the society drive and select the development of technologies. In most cases there is an inextricable feedback from technology to culture and from culture to technology.

Definitions of determinism can get extremely technical. But we can start with the notion of universal causality. This principle is that "every event has a cause," or every event is an effect of some cause or set of causes. Determinism also involves the notion of "same cause, same effect." That is, not just is there some cause for each event, but there is a lawful regularity of the relation of causes and effects. Determinism involves predictability, but is not identifiable with it. For instance, one might have, by accident or by some divine prophetic insight, the ability to prophesy things that happen by knowing the future, but there is no determinism (for this reason some philosophers define prediction as involving a theory, to distinguish it from prophecy). However, if there is determinism, it has in the past been considered to follow in-principle predictability. That is, if there is a causal connection then science, by describing it, can predict the effect. Unfortunately, recent chaos theory seems to decouple determinism and predictability, but part of this depends on what is meant by "in-principle" predictability (see box 6.1).

Furthermore, the notion of human freedom is claimed by many to limit physical determinism with respect to human action (see box 6.2).

There are subclasses of the more general notion of determinism. Various more specific claims as to central causal factors in human life are called determinisms, though proponents of them often do not hold to the extreme of universal determinism. **Technological determinism** is one kind of determinism.

Another kind of determinism is **genetic or biological determinism**. This claims that what we are is wholly determined by our genetic makeup. Walter Gilbert, who tried to patent the human genome and called the sequencing of the human genome "the Holy Grail," claimed he can hold your essence (your genetic sequence) on a CD-ROM and say "Here is a human being; it's me!" (Gilbert, 1996, 96). Lewis Wolpert (1994) claims that we can "compute the embryo" from its DNA code, although other biologists have denied this is possible. Richard Dawkins, in his *The Selfish Gene* (1976), claims we are "robot vehicles blindly programmed to preserve selfish molecules known as genes." This extreme genetic determinism is often used to justify the importance of biotechnology, claiming that we shall be able to totally control the characteristics of agricultural plants and animals as well as ourselves by genetic engineering.

In contrast, the **environmental determinism** of behaviorist psychology in its original form claims that environmental inputs determine all characteristics of the individual. John B. Watson (1878–1958), founder of behaviorism (and later Vice President at J. Walter Thompson advertising agency), claimed:

> Give me a dozen healthy infants, well-formed, and my own specific world to bring them up in and I'll guarantee to take any one at random and train him to become any type of specialist I might select – doctor, lawyer, artist, merchant-chief and, yes, even into beggar-man and thief regardless of his talents, penchants, tendencies, abilities, vocations and race of his ancestors. (Watson, 1926, p. 65)

Four decades later, behaviorist **B. F. Skinner** in his *Walden II* (1966) portrayed a utopian community in which psychologists have totally conditioned the inhabitants to do good, and in which all the residents are perfectly happy. A few years later in his *Beyond Freedom and Dignity* (1971) he contrasted deterministic conditioning of behavior with freedom, dignity, and autonomy, which he rejected as mistaken and old-fashioned notions (see box 6.2).

Box 6.1

Laplacian determinism and its limits

An extreme form of determinism is **Laplacian determinism,** so called because the physicist Pierre Simon de Laplace formulated it in his *Philosophical Essay on Probabilities* (1813). Laplace thought that probabilities were solely a measure of our ignorance, and that every event is precisely, causally determined. Laplace envisions a gigantic intellect or spirit. This spirit is able to know the position and state of motion of every particle in the universe and to perform very complicated and long calculations. This spirit, according to Laplace, is then able to predict every future event, including all human behavior and social changes (under the assumption that human behavior is physically caused). Laplace's spirit could predict what you would be doing tomorrow morning or twenty years from now. God would be an example of such a spirit, but, ironically, Laplace himself was an atheist. When asked by Napoleon why God did not appear in his celestial mechanics, Laplace said, "I have no need of that hypothesis." Laplace thought that he had proven the stability of the solar system. He had not (Hanson, 1964). Since Newton had left it to God to occasionally intervene miraculously to readjust the planets, Laplace's purported proof was the basis of his legendary comment to Napoleon about having no need of God in his celestial mechanics.

One of the great, unsolved problems of mechanics is to find a general solution to the problem of three or more bodies of arbitrary masses in arbitrary initial positions attracting one another by gravitational force. The exact calculation and prediction of the motion of the planets in relation to the sun (including the weaker attraction of the planets for each other) is an example of and motivation for this problem. In the late nineteenth century King Oscar II of Sweden and Norway became disturbed about the possibility that the solar system might fall apart, and, with goading from interested mathematicians, offered a prize in 1889 for a proof of stability, to be awarded on his birthday (Diacu and Holmes, 1996, pp. 23–7). The French mathematician and physicist Jules Henri Poincaré (1854–1912), in his work on this many-body problem of the long-term orbits of the planets, invented aspects of what is now called **chaos theory.** He showed that the solar system might be stable but that stable and unstable orbits are interwoven infinitely close to one another in an infinitely complicated braided pattern. Chaotic systems are deterministic but unpredictable. They can and do arise in classical mechanics and are independent of the Heisenberg considerations (see below). They are mainly caused by non-linear equations (equations that have squares or powers of variables or higher order derivatives) that give rise to "sensitive dependence on initial conditions."

In an ordinary linear system, a slight shift in the initial positions of the objects yields a slight shift of the end positions. However, in a chaotic system, an infinitesimal shift in the initial conditions yields a big shift in the results. Newton's laws of motion are expressed by non-linear equations. Since we cannot measure with infinitesimal accuracy but only with finite accuracy (even forgetting the Heisenberg principle, just sticking to Newton), we cannot predict chaotic systems, even though the math is deterministic.

A common error in the "Science Wars" (see chapter 1) and of romantic or New Age objections to mechanism, even if justified on other grounds (See Chapter 11), is that they assume that since Newton's laws are old fashioned, and that "linear" is boring, Newton's equations are linear. Ironically, it is the newer and more exciting quantum mechanics that is linear (superposition principle), while the older, supposedly less exciting, Newtonian mechanics is non-linear.

Even more severe problems arise for Laplacian determinism from subatomic physics. According to **Heisenberg's uncertainty principle**, it is impossible in principle to simultaneously measure exactly the position and momentum of a subatomic particle (Heisenberg, 1958). Other pairs of variables, such as the energy and the time interval of a process, and particle spin on different axes, also obey the relations. The mathematics of the uncertainty principle is built into the laws of quantum mechanics, which is the best theory we have of the structure of matter. According to Heisenberg's principle, even Laplace's demon couldn't know the simultaneous position and motion of one particle, let alone of all of them. This is not a matter of our inaccuracy of measurement, but is built right into the equations of the theory. The operators differ in the value of the product, depending on the order in which one multiplies them (to use mathematical jargon, they are non-commutative). The difference between AB and BA, for these operators, equals the limit of accuracy, related to a fundamental constant of physics (Planck's constant). Despite Heisenberg uncertainty at the atomic level, we can have statistical determinism in many systems, where we cannot predict exactly, but can predict within a range of error. Surprisingly, the teenage Heisenberg first learned of the idea of geometric structures of atoms from reading Plato, before he was disappointed by the crude materialism of the tinker-toy models in his chemistry text. His opposition to materialism was in part tied to the fact that he was defending his laboratory against Marxist revolutionaries, and read Plato's theory of the creation of the universe in Greek for relaxation on the military college roof at lunchtime (Heisenberg, 1971, pp. 7–8). Heisenberg later claimed to have been stimulated by Plato's concept of the receptacle (matrix, or mother), a spatial principle leading to fuzziness or inexactness when the perfect forms shape real objects in space. Despite this early interest in Plato's metaphysics, when Heisenberg first presented the mathematical skeleton of his quantum mechanics, he understood it in a strictly instrumentalist and positivist manner, claiming that the mathematics was merely a tool for making predictions, not a picture of reality (see chapter 1). Later in life, Heisenberg understood his theory in terms of Aristotle's theory of potentiality and actuality. For Heisenberg, the abstract, mathematical states are objective, but potential in nature, while the physical observations are subjective but actual, turning the usual notions of reality on their head.

Many people think that randomness occurs in the behavior of individuals (perhaps due to free will), but that statistical regularity of populations is true in the social sciences, such as sociology. The social sciences often in practice use statistics when they are attempting to be rigorously mathematical. Laplacian determinism would claim that probability is just a matter of our ignorance, our lack of exact knowledge. Chaos theory and quantum mechanical indeterminacy suggest that for complicated systems of many particles, such as a human being or a society, such statistical methods may be necessary, and not replaceable with exact methods even in principle. It is perhaps on analogy to this situation that the notion of statistical determinism in the social sciences, called by Mesthene and Heilbroner "soft determinism," is sometimes conceived.

Though the Heisenberg principle applies to the subatomic level, its operation is usually effectively so minuscule as to be unnoticeable in larger, many-particle systems. However, there are significant places where it has been suggested the subatomic indeterminacy may be "magnified" to have effects on the macroscopic level. These include certain biological mutations involving small changes (of a single atomic bond) of the genetic material that controls biological heredity (Stamos, 2001). Another, more speculative, example is in the firing of individual tiny dendrites or fibers in a neuron in the brain, where the chemical changes are on a small enough scale for the Heisenberg principle to be applicable (Eccles, 1953, chapter 8). More recently, physicist Roger Penrose has suggested that collapse of the quantum wave function in the microtubules of the brain is responsible for the non-mechanistic aspect of thought (Penrose, 1994, chapter 7).

Thus determinism has been challenged within physics from at least two directions. Chaos theory suggests that even the in-theory, deterministic mathematics of Newtonian mechanics can yield situations in which practical prediction is impossible. Quantum mechanics produces an even stronger, in principle, objection to universal determinism, in that the indeterminism is built right into the mathematical core of the theory. Free will is an older, more concretely human problem in relation to determinism (see box 6.2).

Box 6.2

Freedom and determinism

Whether one thinks that humans are enmeshed in a deterministic technological system that controls their actions, or whether one thinks that humans freely construct their technology and society, implicit positions on the problem of freedom versus determinism are assumed.

One of the traditional philosophical problems is that of freedom versus determinism. In the Middle Ages the conflict was formulated in relation to the idea that God, who was omnipotent (all-powerful) and omniscient (all knowing), could foresee and control all, but let Adam sin. If God knew that Adam and Eve would fall when they were created, were they really free? Predestination (the doctrine that God predetermined at creation who will be saved or damned) also conflicts with freedom. Figures such as St Augustine (354–430) and the mathematician/philosopher Gottfried Leibniz (1646–1716) claimed to have reconciled these conflicts.

With the rise of scientific laws and determinist ideas the conflict with freedom took a new form. If everything we do is physically causally determined, can we be free? (See the discussion of Laplacian determinism in box 6.1.) In this dispute determinists say no, while so-called **libertarians**, who are believers in metaphysically free will (not the same as political libertarians, who support minimal government), say yes, we really are free. There are two senses of freedom here. One is the **contra-causal** sense, the one that conflicts with determinism. In this sense of freedom, acts of free will counter physical causes. That is, acts of free will are said to somehow break or go against the chains of physical, causal determinism. The other is the sense of **freedom as responsibility**, which need not conflict with physical determinism.

Compatibilists attempt to reconcile freedom and determinism. A way to accomplish this is to say that we are responsible for our acts even though we are determined. One can try to claim that responsibility has nothing to do with the contra-causal sense of freedom. One way of claiming that freedom and determinism are compatible is to say that free acts are simply the ones that issue from us in some sense. Even though all acts are ultimately determined, we can distinguish between acts that, in some sense, issue from us, and acts that are produced by external physical causes (such as being blown off a roof and falling on someone) or external human coercion (such as being threatened at gunpoint). This is an approach commonly taken in various forms by British empiricist philosophers such as John Locke (1632–1704) and John Stuart Mill (1806–73).

Another way to combine freedom and determinism is to claim that the *physical* world is completely deterministic, but that the mind or soul is a different sort of non-physical stuff and is free. This is the **dualism** of the seventeenth-century philosopher-mathematician René Descartes (1596–1650). This is a substance dualism, in that Descartes claimed that there are two kinds of entities (substances), material substance and mental substance. That is, matter or physical stuff and mind or mental stuff are totally distinct from one another. The problem that arises from this is: if these substances are so different in nature, how can one affect the other, how can the non-physical mind affect the body? This is called the **mind–body problem**.

Because of its difficulties few philosophers hold this **Cartesian dualism** or **causal interactionism** today. Nevertheless, several Nobel Prize winning brain scientists, including Sir Charles Sherrington (1857–1952), Sir John Eccles of Australia (1903–97), Wilder Penfield of Canada (1891–1976), and Richard Sperry of the USA (1913–94), held this position in the late twentieth century. Their dualism cannot be dismissed as based on ignorance of brain science, since they were its leaders. It arose from the mystery of how experienced consciousness is related to the physical brain.

(As mentioned concerning the Heisenberg uncertainty principle, Sir John Eccles, like the astrophysicist Sir Arthur Eddington (1934, pp. 88–9, 302–3), appealed to the effect of quantum indeterminacy on the brain cells. However, even if this could yield random actions, we wouldn't necessarily consider them freely chosen actions.)

Another, and somewhat less problematical, way of reconciling freedom and determinism is in terms of **two standpoints** theory, the solution of the eighteenth-century German philosopher Immanuel Kant. Kant claims each individual has a dignity that is of infinite worth. Kant believes that we are genuinely free. He also claims that the world is completely deterministic. From the point of view of knowledge and science we structure the world in terms of deterministic laws, and can only understand it in these terms. We search for laws and see things in terms of causal laws. However, from the point of view of morality we consider ourselves as free. Moral acts are based on a moral law that we freely legislate to ourselves. Acting in terms of the moral law, we freely choose. Thus Kant reconciles freedom and determinism by a kind of dualism of standpoints or perspectives, rather than a dualism of soul and body. That is, we can describe human behavior from the standpoint of science and causal laws, and we can describe human behavior in terms of moral responsibility for our acts. These two approaches are different but do not conflict. They are mutually applicable. In the twentieth century there have been variations on this sort of reconciliation of freedom and determinism based on different uses of language that have been popular. One such approach distinguishes principled rational reasons for acting from causes of acting and claims that the language of human reasons and the language of physical causes are quite distinct. Other philosophers disagree with this and claim that reasons do function as causes of action, thus undermining the two languages approach to reconciling freedom and determinism.

The conflict of the claims of determinism and freedom has been one of the great issues of philosophy since the Christian writers of the Roman Empire. The development of the concept of laws of nature set the problem within a new framework, but did not eliminate most of the traditional issues. The interplay of freedom and determinism occurs within society and throughout the history of technology. Some attempt reconciliation of the competing claims in terms of a "soft determinism." Others found their accounts on the total freedom of the active subject in social constructivism. Still others eliminate the role of freedom for the impact of events on the passive subject in some forms of structuralism. Yet others claim to solve the dilemma by eliminating the subject altogether, as in some forms of postmodernism. Various positions on freedom and determinism lie in the background of debates about the nature of humans in a technological society.

Biological determinists sometimes say we are lumbering robots controlled by our genes. Behavioral, environmental conditioners such as Skinner say we are controlled by the inputs from our environment. Technological determinists and economic determinists claim that technique or the economic system determines political and cultural phenomena such as art and religion.

Some theorists of technology, such as the founder of the first university based technology and society program, **Emmanuel Mesthene**, and the economist **Robert Heilbroner**, use the term **soft determinism**. This is to be contrasted with hard determinism. They get this term from the 1900 philosopher and psychologist William James. James meant by soft determinism compatibilism (see box 6.2). (By hard determinism, James meant a determinism that excluded freedom.) But Heilbroner and Mesthene use it to mean something more like **statistical determinism** – that there is freedom, but that there also are larger statistical trends. French sociologist **Emile Durkheim** (1858–1917), in his book *Suicide* (1897), used this sort of notion to claim that there were social regularities concerning rates of suicide even if one couldn't predict whether a given individual would commit suicide. Catholics committed suicide less than Protestants, peasants less than city dwellers.

Marx and Heilbroner on Technological Determinism

Karl Marx in the preface to *A Contribution to a Critique of Political Economy* (1859) presented in a short passage the briefest, clearest summary of his theory of social change. Marx may have made it so neutral and objective to get by the censor. This passage, unlike Marx's earlier humanistic discussion of alienation, which was not rediscovered for seven decades, was soon published and well known to socialists and communists. It was highly influential on the later socialist and communist movements. It influenced later technological determinism not only of Marxists, but also of twentieth-century neoconservative technocrats (see chapter 3).

TECHNOLOGICAL DETERMINISM

Marx distinguishes between the **foundation** (**base**) and the **superstructure** in an architectural or civil engineering metaphor. The foundation is economic in a broader institutional sense than contemporary microeconomics, and contains two major components. The first, the **forces of production**, include energy sources, including human labor, and technology. Emphasis on this part of Marx's theory of society legitimates technological determinism. The economic base also includes the **relations of production**, which are the power relations within production, such as slave-owner versus slave in slave society, lord versus peasant in feudal society, and owner versus wageworker in capitalist society.

The economic base determines or causes the superstructure, which includes legal relations (which codify the relations of production), politics, and more ideational realms such as religion, philosophy, and art. Marx claims that the religion and philosophy of a society are determined by the economic foundation, those forces and relations of production. Notice that the place of science is ambiguous in this. It "seems to float between base and superstructure" (Mills, 1962, p. 105). Via technology, science is a force of production. At the same time, influenced by theories of the nature of reality (metaphysics) and of method, in turn influenced by social worldviews, science is an intellectual ideology like philosophy and religion.

Social change occurs because the foundation changes more rapidly than the superstructure. Sociologist William F. Ogburn (1886–1959) later called this "cultural lag" (Ogburn, 1922, p. 196). The political and religious superstructure becomes dysfunctional to the economic foundation. For instance, in eighteenth-century France there was a growing capitalist, factory technology. Nobles and priests ruled the country, while a new class of capitalists was growing, but not a part of the government. The laws were medieval, while capitalist property was growing. Eventually, the whole superstructure collapsed and reorganized to adapt to the base.

Marx used this scheme to explain the French Revolution of 1789–1815. Marx distinguishes between the real economic forces and the ideals of religion and politics in terms of which people fought out the conflict. The French revolutionaries thought they were reviving the Roman Republic and Empire. The English Puritans of the Puritan Revolution of the 1640s thought they were reviving the society of the biblical Old Testament. Both French and British revolutionaries ended up creating capitalism behind their own backs (Marx, 1852, pp. 5–17). This is what Marx meant by ideology. It is for him the false consciousness in politics, religion, and philosophy that serves the ruling class. Emphasis on forces of production (energy sources) leads to technological determinism such as the USSR version of Marxism supported. On the other hand, emphasis on relations of production, claiming that class power relations determine which technologies get used, was supported by Maoist Chinese Marxism as well as many late twentieth-century non-communist Marxist writers.

Robert Heilbroner is an American institutional economist who wrote the very popular history of economic thought *The Worldly Philosophers* (1952). At the opening of his essay "Do machines make history?" Heilbroner (1967, p. 335) quotes Marx: "The hand mill gives you the feudal lord. The steam mill gives you the modern capitalist." Heilbroner notes that the models of science as cumulative and linear, and of technology as applied science, support the idea of technological determinism. That is, if there is a single necessary path for science, and technology is the straightforward application of that science, then there is a single linear path for the development of technology. The direction of technology is not swayed by other cultural factors.

Heilbroner cites **Robert K. Merton** (1910–2003), who summarized the frequency of multiple **simultaneous discoveries**. Simultaneous discoveries or "multiples" are independent discoveries by several people at the same time. Examples include: Newton and Leibniz discovering the mathematics of rates of change (the differential and integral calculus); Darwin and Wallace both discovering natural selection; and some six people independently hitting upon the principle of the conservation of energy. This frequency of simultaneous discovery suggests that science isn't just an arbitrary product of geniuses, but that ideas develop in sequence. (Of course, someone supporting the cultural determination of science could claim that the shared social situation accounts for the multiple discovery, such as both Darwin and Wallace reading Malthus on population and modeling biology on capitalist competition.) Heilbroner examines "inventions ahead of their time," such as the ancient Greek steam engine in Alexandria, Egypt, and Babbage's design for a huge gear and board programmable computer with the basic architecture of a modern computer in 1830s England. A technological determinist explanation of this is that metallurgy wasn't up to making iron that would resist the strong pressures needed to build a large steam engine. (Though not noted by Heilbroner, a similar explanation partially accounts for the non-appearance of Babbage's computer: that metalworking wasn't up to building accurate enough gears for Babbage to avoid analogue error multiplication.) In contrast, a Marxist "relations of production" explanation might be that Greek slave-owners wouldn't like their galley slaves being replaced by a steamboat or mining slaves replaced by a steam pump. The machines would make the majority of the slaves, and hence the present wealth of the slave-owner, superfluous. This assumes that the slave-owners were insufficiently far sighted to envisage turning themselves into capitalists, becoming owners of the machines, and dispensing with the majority of their slave property in order to acquire machine capital.

Modern western technocrats and theorists of post-industrial society reject Marx's claim that socialism and communism would replace capitalism, but they do hold to a technological determinism of the sort described above. They claim that because of electronics and communication technology society is evolving from an industrial society centered on manufacture to a post-industrial society centered on information and services. This transition is similar to the earlier replacement of agricultural society by industrial society. The changed technology brings changes in work and in the politics of the society. For technocrats the post-industrial society shifts power from traditional capitalists to technocrats. Different versions of this theory claim that owner-capitalists are replaced by managers as the controllers of industry and/or that in government politicians cede power to technocrats, including technologists and social scientists who supply the information on which politicians act and the options among which they decide.

Forms of Information: A Communications Version of Technological Determnism

Modern technocrats and theorists of what has been variously characterized as the information society, service economy, post-industrial society, and "second industrial revolution" emphasize information rather than energy in their technological determinist accounts of contemporary society (see chapter 3). **Marshall McLuhan**, whose most famous slogan is "the medium is the message," developed a theory of communication media and their effect on human consciousness and culture.

McLuhan's account resembles Marx's broad historical sequence scheme in one respect. It involves an early communal stage of society (in Marx primitive communism, and in McLuhan oral society), a later alienated one (in Marx capitalism, and in McLuhan print culture), and a return to the happy communal stage again (in Marx communism, and in McLuhan television-centered society) (Quinton, 1967). Both McLuhan's and Marx's schemes resemble the biblical one of the Garden of Eden, initial innocence, a fall into sin and expulsion, then eventual salvation.

In archaic society or non-civilized society communication is oral. Everything has to be memorized and handed down orally. Chanting, poetry, song, and rhythm help people to remember long epics or genealogies. The bard or chanter speaks surrounded by the local community. With writing this no longer becomes necessary. Writing is visual, not oral. Reading is private, not communal like a recitation. Writing separates the author's personal presence from the work. Writer and reader are separated from one another, unlike the oral storyteller and communal audience.

Despite the separation of author from physical printed work, print allows the author to become important. In the oral tradition, although the bard was greatly respected, the bard was not the source of the story (even if good bards considerably elaborated on and improved it). The story was supposedly a traditional myth. With the printed work, the author is generally thought to be the source of the text.

Engineering diagrams and anatomical drawings can be accurately reproduced in printed works. Multiple copies and cheap printings allow knowledge to be available outside the monastery or castle. The printing of the Bible led to the Reformation, as the priest no longer had monopoly on the text or interpretation.

McLuhan's account is not unique. A number of scholars in history, the classics, and rhetoric developed the contrast between oral and literate culture. Writers in communications theory (Ong, 1958) and in classics (Havelock, 1963) had developed similar ideas, but without the grandiose sweep and popularization of McLuhan. McLuhan extended the series of contrasts of communication media to include twentieth-century television.

According to McLuhan, the new electronic media changed human experience and culture once again. McLuhan strangely calls radio a "hot" or emotional medium. Radio brings back the voice and personality of the speaker, which had been eliminated in print. (Many critics rejected McLuhan's treatment of radio as "hot" when his book first appeared, but the development of angry political radio talk show hosts and angry audiences of what has been called "hate radio," as well as an attempt at liberal imitation of this discourse on Air America, suggest McLuhan was on to something.) Television brings in visual context and experience, which McLuhan claims makes television a "cool" medium. Television makes people more primitive, into what McLuhan called "the global village." For McLuhan, while print corresponded to Marx's alienation, television will save us by returning us to a primitive communism or Garden of Eden.

Some German theorists of media, such as "critical theorists" T. W. Adorno and M. Horkheimer, exposed to Hitler's dictatorship and propaganda followed by exile in Hollywood, were less favorable than McLuhan to the electronic media. They went as far as to claim a resemblance between fascism and twentieth-century mass media such as cartoons and television. Hitler, significantly, was the first politician to make highly successful use of the radio (and the first political campaigner to make use of the airplane). The view is given support by more recent developments, such as the centralization of modern television and radio owned by half a dozen major corporations (Bagdigian, 2004), the emotional nature of television news, the angry hosts of the television news commentary, and the sometimes obscene hosts of news-oriented comedy shows encourage a kind of reactionary emotionalism. The centralization of television can allow a kind of mass brainwashing by propaganda. Adorno, in his "How to watch television," called television "reverse psychoanalysis" (Adorno, 1998). Rather than making us consciously aware of our primal urges and neuroses, as psychoanalysis is supposed to do, television makes us become unconscious and infantile. McLuhan and Adorno agree on the resulting primitivism, but differ in their evaluations of it.

Probably the truth lies somewhere far from McLuhan's uncritical praise but also considerably short of the critical theorists' blanket condemnation. Studies of the extent to which cartoon violence on television encourages or conditions actual violence in children who watch it have been criticized and debated, because of the artificial environment of many of the experiments, but seem to show an effect. However one evaluates the results of television watching, the claim that television technology has affected the psychology and culture of its viewers cannot be doubted.

The Internet developed after McLuhan's time but lends itself to analysis of the effect of communication media on experience and culture. The Internet is decentralized and contrasts with television in this respect. In addition, it includes print along with music and videos. The Internet is interactive, while television is one-directional from a broadcasting center that depends on huge financing. At the turn of the twenty-first century a huge consolidation of US media has occurred, in which the major television networks are owned in large part by a half dozen or so large corporations, including ones such as Westinghouse and General Electric with many other interests, such as nuclear power (Bagdigian, 2004). If the Internet remains and increases in decentralization and is not captured by commercial firms, service providers, and cable companies, it may counter the centralization and passivity that television encourages. However, increased fees for access to virtual information and censorship of it may decrease the democratic and anarchistic aspect of the Internet. There has been privatization and commercialization of some formerly free services and resources on the net, but free downloads, individual websites, and the explosion of blogs have maintained much of the net's libertarian character.

Writers such as Dreyfus (1999), ingeniously using the categories of the aesthetic way of life from the Danish existentialist Kierkegaard, and, in a more critical but less negative way, Borgmann (1995, 1999) (see chapter 5), have criticized the Internet as devaluing education with mindless surfing, and alienating the individual by swamping her with an overload of meaningless information. Like Adorno's horrified critique of radio and television, these criticisms have a grain of truth, but tend to underplay the positive aspects of the net. The Internet has vastly expanded accessibility to technical information for people at non-elite educational institutions and academics in developing nations, though it is accessible only to a minority in the developing nations. Even in the affluent nations, not all neighborhoods have an exit on the "information highway." It also has the potential to increase democratization, even if to a lesser extent than claimed by its most utopian partisans (see box 6.3).

Cultural Determination of Technology

In recent decades technological determinism has come in for extensive criticism from writers on technology. To criticize technological determinism authors have presented examples that are claimed to show that society has effects on the direction of or acceptance and rejection of technology. One way of doing this is to show that alternative directions for the development of technology were available and a socially influenced choice was made. Often this is difficult to do, since once a technology has been settled on it acts as a constraint on further directions of development. The technology then appears in retrospect to be inevitable, and this supports belief in technological determinism.

One version of Marxism (the base and superstructure model, in which the technological base determines the political and cultural superstructure) has been taken by many as a classic formulation of technological determinism. However, Marxists have given examples of social determination of technology. One is that in a society based largely on slave labor there is not the

Box 6.3

Postmodernism and the mass media

Some have claimed that the approaches and theses of so-called **postmodernism** may be a product of the growth of the mass media and of the Internet. This implies that the information society is the source of postmodernism. This is an example of the "forms of information" version of technological determinism, in which different forms of communication media replace the Marxian forces of production as the determining factors of culture. (see McLuhan's theory of media as an earlier example of such informational determinism). As one moves from oral, to print, to radio and television, to the Internet, new forms of worldviews and theories are caused by the new forms of media.

Postmodernism is a broad and diverse movement that has been influential in the recent decades in the humanities and social sciences, including science and technology studies. Some major features of postmodernism are: (a) emphasis on the language-structured nature of our grasp of reality; (b) denial of the possibility of "grand theory" in terms of general metaphysical, philosophical theories of the nature of reality, or grand social theories and "grand narratives" that claim to account for the whole sweep of history; (c) denial of a unified self that is the center of our understanding of the world or a basis for politics; (d) denial of unifying essences or natures of things; and (e) denial of overall human progress. Clearly postmodernism rejects some of the assumptions of classical theories of figures such as Plato and Aristotle, such as the reality of essences and the possibility of a general metaphysical system. Postmodernism also rejects the views of the rationalists of the early modern period, with their emphasis on the capabilities of rational thought in science and philosophy to grasp the nature of reality. Finally, postmodernism rejects eighteenth-, nineteenth-, and early twentieth-century theories such as the Enlightenment, positivism, and Marxism, with their belief in science and human progress.

Lyotard (1979), in a central manifesto of postmodernism, rejects "grand narratives" of history of the sort produced in the nineteenth century by Comte and the positivists, Marx and Marxists, or Spencerian evolutionists. Lyotard opens his account by referring to computerized societies and the "post-industrial age." According to him the new electronic era discredits and makes obsolete the narratives of progress, as well as those of the growth of state regulation.

Postmodernism, among other things, emphasizes the role of language and writing in the broadest sense. One of the features of postmodernism is the denial of a separate objective reality distinct from the symbolic or linguistic presentations of it. The rejection of traditional objective concepts of reality and science parallels the growth of the development of and speculation about virtual reality in computer science and science fiction. For example, one very idealistic Internet manifesto "A declaration of independence of cyberspace," by John Perry Barlow of the Electronic Frontier Foundation (1996), claimed that denizens of the Internet were independent of physical environment: "you weary giants of flesh and steel, I come from Cyberspace, the new home of Mind. . . . Ours is a world that is both everywhere and nowhere, but it is not where bodies live. . . . Your legal concepts . . . do not apply to us. They are based on matter. There is no matter here" (Barlow, 1996; Ross, 1998). A major postmodern sociologist, almost as if to unknowingly satirize the subjectivism of postmodernism, has claimed that the best way to deal with the first Gulf War is to deny that it ever happened (Baudrillard, 1995; see also Norris, 1992).

Postmodernism denies the possibility not only of general metaphysical systems accounting for the universe as a whole but also of general social theories such as Marxism. Postmodernism typically emphasizes the fragmentation and lack of any unifying structure to society. It then goes on to deny that universal theories (and "grand narratives" of history such as the theories of progress of St Simon, Comte, Marx, and many others) are possible as adequate accounts of society. The global interconnectedness of television and the Internet undermines traditional nationalities and local cultures and in turn generates reactions, such as American Christian, Islamic, Jewish, and Hindu "fundamentalisms," as well as local separatist movements in Scotland, France, Indonesia, and the Philippines. According to some postmodernist accounts, the weakening of national governments (though counteracted by wars and the military) can strengthen local secession at the same time that the economy and culture is globalized. The Internet facilitates international communication in social movements. For instance, the Zapatista peasant rebels in Chiapas, Mexico, could get international support via the Internet, through the ingenuity in communication of their university-educated spokesperson "Subcommander Marcos." Anti-globalization protestors who demonstrated at World Trade Organization, International Monetary Fund, World Bank, and World Economic Forum gatherings communicated via the Internet to organize protests opposing the very economic globalization that the new telecommunications technology has played a role in facilitating.

drive to develop labor saving devices found in a society based on salaried labor. This was briefly mentioned above. The ancient Greeks had scientific knowledge superior to the societies that succeeded them up to the seventeenth century. Nonetheless, the Greeks did not develop machine technology. The Greeks had knowledge of gears and in Alexandria a model steam engine was invented. Nevertheless, one does not find these innovations applied to practical devices such as pumps or steamboats. These arose only in the early modern period. The claim is made that once wage labor became the dominant form of work there was a drive to save on wages by multiplying the strength and speed of the individual worker through mechanical devices.

During the Middle Ages peasants who paid their lord a fraction of their crop had replaced slave labor. In monasteries there was a desire to save labor so that monks could devote themselves to their religious duties. Hence in monasteries there was a great deal of invention of mechanical devices. Religious interests also affected invention. The mechanical clock was developed to time the prayers that were to be said every few hours, especially those at night (Mumford, 1934).

In the twentieth century there were cases in which alternative solutions to a technological problem were available, but one was chosen for social or political reasons. David Noble (1984) claims that the numerically controlled lathe is an example of this. An earlier, alternative approach was a lathe that recorded the motions of a skilled worker on tape. The tape could be played back to produce the pattern without the manual guidance of the worker. This latter playback technology gave the lathe operator more control over the process. New and better recordings could be made to replace the original one. However, the numerically controlled lathe came to replace the playback device. Noble argues persuasively that the early numerically controlled lathes were not more accurate or efficient than the playback lathe. In fact, the numerically controlled lathes often malfunctioned and had to be adjusted or overridden by the workers. Noble argues that management preferred the numerical control technology because it put control of the process in the hands of the engineers and managers, not of the lathe operators. The desire for control on the factory floor trumped efficiency. The technology developed was not the inevitable result of the quest for engineering speed and efficiency. Instead, it was a product of the desire to maintain greater control over the workers.

Social constructionist students of technology also criticize technological determinism (see chapter 12). Social constructionists claim that many interest groups influence the development of a technology. Various groups prefer various alternatives. In Bijker's (1995) study of the bicycle, different early designs maximized speed at the expense of safety or maximized safety with a sacrifice of speed. The early bicycles with a huge front wheel were built for speed but were harder to balance and led to a rider's fall from a greater height if they tipped. The early versions of the smaller wheeled bicycles were safer but less fast. Different users, those interested in racing and those interested in leisurely recreational touring, preferred different designs. The big wheeled, fast, but unstable models attracted primarily athletic young men. The resulting standard bicycle was a product of the pushes and pulls of the divergent interest groups.

Andrew Feenberg (1995, pp. 144–54) has described how consumers have actively modified technological designs or plans because of their interests. One example of this is the French minitel. Originally this device, introduced by the telephone company, was meant primarily as a means to access information: weather forecasts, stock market results, and news headlines. Users hacked the device to modify it into a communication or messaging device. Eventually this became the main use of the device, far eclipsing the information access function for which it had been designed.

Another example that Feenberg (1995, pp. 96–120) uses is the influence of AIDS patients on drug trial protocols. Since untreated AIDS is fatal, the sort of double blind medical tests usually used were, in fact, sacrificing the lives of those patients who were given placebos and told that they were getting AIDS drugs. Patients rebelled against this situation. At first the medical researchers saw the resistance as anti-scientific and irrational. However, eventually researchers came to work with the patient advocacy groups. The latter became extremely well informed scientifically and influenced the drug testing procedures. Feenberg notes that the overly strict identification of medicine with pure "science" neglects the normative elements of medical treatment procedures. The attention and study given to experimental subjects often offers the kind of personal care that chronic patients miss and value.

Arnold Pacey (1983, pp. 1–3) discusses how the snowmobile used by recreational users in southern Canada and the northern USA is basically the same device as used by Inuits in far northern Canada. However, the routine of maintenance and use is very different. The recreational user in the populated southern areas generally does not go far from home. If a breakdown occurs there are generally phone booths and gas stations to deal with the problem. For the First Nations or Inuit user the snowmobile is a part of survival in hunting and other activities. In the far north breakdown far from any repair facilities could be deadly. The snowmobiles must be kept in shape for long journeys, and carry extra gasoline. Some Native Americans in the far north keep the snowmobile in their lodging to keep it warm. Although the technological modifications for the cold north may be minimal, the culture of maintenance leads to very different use of the technology.

The science and technology studies community, with its penchant for social constructionism, generally rejects technological determinism as outdated. However, recent technology studies at times falls into the opposite extreme of a thoroughgoing cultural determinism. This is partly due to the fact that the social study of technology modeled itself on the field of sociology of scientific knowledge (SSK). Theory is important in physical science and the traditional philosophy of science. Because of this, SSK tended to emphasize theory and the social construction of ideas rather than of technical apparatus. Despite the development of instrumental realism in philosophy (see chapter 1) and the growing emphasis on experiment in SSK, technology studies, as a latecomer, based itself on the earlier SSK.

Bruno Latour (1992, 1993) has criticized the overemphasis in social constructivist sociology on culture producing technical objects and the underemphasis on the role of objects in producing culture. Latour emphasizes that neither nature nor culture is primary, that objects should be treated as "actants" symmetrically with people, and that the starting point for analysis should be the nature–culture hybrid. Donna Haraway, with her use of the cyborg as symbol of the erasure of the nature–culture dichotomy, and her more recent use of domesticated dogs as an example of the same, is equally concerned with this issue. Andrew Pickering, likewise, has emphasized with his "mangle of practice" that physical objects and regularities, as well as political and cultural considerations, act as constraints on the activity of scientists and technologists (Pickering, 1995).

Study questions

1 Give an example of a technological determinist account of the effects of some invention not discussed in the chapter.
2 Do you think that the existence of human free will shows technological determinism to be false?
3 Does a "soft determinism" or statistical determinism hold despite human free will?
4 Do "unintended consequences of human action" lead to a kind of technological determinism?

7

Autonomous Technology

The claim that technology is autonomous is the claim that technology is independent of human control or decision. Technology is claimed to have a logic of its own or, more metaphorically, technology has a life of its own. The notion of autonomous technology is associated with the French writer Jacques Ellul and his *Technological Society*.

The claim seems paradoxical, as human beings invent, market, and use technology. However, Ellul notes that the various groups of people that would seem to have control of technology do not do so. The technologists and engineers who develop technology lack understanding of the social impact of technology and are often naive about the means of controlling it. (Ellul claims that brilliant physicists, such as Einstein, Bohr, and Oppenheimer, whose work is the basis of nuclear weapons, were often extremely naive about the possibilities of disarmament and the sharing of American nuclear knowledge with the Soviet Union.) Ellul is also contemptuous of the usually benign portrayals of future society's use of technologies that technical experts project.

After-dinner speeches concerning the future of technology traditionally spoke of "man choosing" technologies, as if the choice was unconstrained and mankind as whole rather than powerful interest groups made the choice. The politicians and businesspeople who support research lack understanding of the technical aspect of technology. Certainly Ellul is correct that people, whether engineers, businesspeople, or politicians, do not grasp the consequences of the technologies that they develop or advocate. While the technologists are generally ignorant and naive about the social and political issues surrounding a technology and the politicians are often abysmally ignorant of the workings of the technology itself, Ellul claims that the public is ignorant of both the technical and the social aspects of the technology. Ellul further claims that the religious leaders who are supposed to deal with the value issues of society's use of the technology are ignorant of the technical and social issues, while no one listens to the philosophers who evaluate technologies. Autonomous technology is a special case of the unintended consequences of human action that Max Weber discussed.

A common set of beliefs about scientific method and the nature of technology support the autonomy of technology (Heilbroner has used these claims to support technological determinism, but they are even more relevant to autonomous technology). If one thinks of scientific method as fairly mechanical and cut and dried (such as the Baconian inductive method), and thinks that the nature of reality, not human interpretations and theories, is what determines the sequence of scientific discoveries, then the sequence of scientific discoveries is predetermined and linear. If technology is simply applied science, then the necessary logic of the linear sequence of scientific discoveries predetermines the linear sequence of technological applications. Very rarely is a technology withdrawn once it has been introduced. Japan withdrew firearms because they undermined the heroic and knightly samurai ethic and the feudal social structure of which it was the pinnacle. However, eventually Japan reintroduced them.

While Japan was isolated from the rest of the world the regime was able to suppress firearms, but when in the nineteenth century foreign powers were seen as a threat, Japan's rulers embraced the gun and reinstituted arms production (Perrin, 1979). If people rarely or never reject a technology, then the sequence of technological inventions that society accepts unfolds automatically from the nature of the world and the nature of scientific method. Thus technology can be claimed to have a logic of its own, independent of human desires.

Another feature of technology that supports the autonomous technology thesis is its tendency to spawn more technology. Ellul notes, as have others, that technologies constantly produce unanticipated problems. Generally the solution to these problems is more technology, not the rejection of the technology.

Moreover, society tends to adjust to the technology, rather than adjusting the technology to society, claims Ellul. Ellul himself prefers the term "technique" to technology and identifies technique with a set of means–end relationships and rules for achieving maximum efficiency in the adjustment of means to ends (without the ultimate ends ever being examined). Ellul's characterization of technique fits with the definition of technology as rules.

However, the system definition of technology, including human organization and the marketing and maintenance of technological devices, can support the autonomous technology thesis. In what Langdon Winner (1977) calls "reverse adaptation" the technological system, particularly its social components, adapts the society to technology rather than vice versa. The marketing of technology persuades the public, through advertising of particular devices or propaganda for the acceptance of technologies in general, to accept new technologies (Ellul has written extensively on propaganda itself as a technology). This claim denies that technologies are simply freely chosen by the public in the market place but emphasizes the ability of advertising to sell the technologies to the public. In the case of large-scale technologies, government institutions, such as the nuclear weapons program or the space program, influence governments to support and fund a technology. Once large investments of money and commitment have been made in a technological project, it is more and more difficult to simply drop it, even if problems and difficulties arise. (The Vietnam War is a prime example of this sort of situation. The larger the commitment one has made, the more difficult it is to withdraw. A possible exception to this was the superconducting supercollider (SSC), abandoned after two billion dollars was spent, but this project did not engage the nationwide economic and political commitments that a war does.)

Similarly, large corporations lobby the government to support the production and public acceptance of the technologies that they manufacture. An example of government and corporate pressure to support the nuclear power industry is the US Price Anderson Act, which limits the insurance that would be paid in the event of a nuclear catastrophe. If nuclear power companies had to pay the full insurance rates to cover major nuclear accidents the cost of insurance would be prohibitive, so an act was passed to limit insurance payments to a tiny fraction of the cost of such an accident. Thus, according to the doctrine of reverse adaptation, large-scale technologies tied to powerful institutions and interest groups ride roughshod over any social resistance.

Langdon Winner in his elaboration of the notion of autonomous technology emphasizes that the tool model of technology misleads people about modern technological systems and supports naive and self-serving justifications of the present state and directions of technology. The notion that "man controls technology" not only assumes a kind of collective person but treats modern technology as a tool that could be manipulated by an individual. An individual uses the tool for a particular purpose, to reach a particular goal. The technological system, on the other hand, at best has its end products utilized by consumers, but is not really used, in the sense of guided toward a particular purpose by anyone. The consumer does not originate, maintain, or understand the complex technology or complex socio-technology of the system. However, even the inventors, engineers, maintenance people, businesspeople, bureaucrats, politicians, and others involved in the system lack overall intellectual grasp or strategic control of the system. They control as well as understand only a small fragment of the system.

Indeed, discussing a technological system in terms of its manipulation to reach a goal or serve a purpose is misleading, according to Ellul and Winner. The goals of technology are fine sounding – prosperity, progress, happiness, freedom – but often in practice rather empty of content. The goals become abstract and empty, while the means become ever more complex and re-fined. No one questions goals such as progress and wealth, but their content is left vague. The focus is on the means of development. Through reverse adaptation and similar mechanisms the goals themselves are given content to fit the means available.

Winner suggests that autonomous technology, far from being carefully controlled and directed by a technological elite, is really ruled or controlled by no one. The complexity of the system, with its constraints and imperat-ives, guides everyone serving it to behave in the appropriate manners and carry out the appropriate rules. Even the state does not play the all-powerful central planning role that many technocratic theories describe or advocate. The various large technological systems follow their own rules and at times conflict with one another, but are not truly controlled by the state.

Autonomous technology, with its emphasis on the incomprehensible com-plexity of modern technological systems, rejects one of the theses of the precursors of the social construction of technology notion. Vico and Hobbes claimed that we truly know that which we make or construct (see box 12.1). However, much that we make develops a complexity that outruns our abil-ity to understand it. Many computer programs are so complex (and if they involve learning or natural selection become ever more complex) that we cannot grasp exactly how they operate. This is true of much artificial intel-ligence utilizing neural networks or evolutionary computing.

Winner draws on writers such as Arendt (1964) on Adolph Eichmann, or Albert Speer's own justifications, to claim that individual responsibility is eroded and eliminated in the technological system. The Nazi bureaucrat Eichmann claimed that he was just doing his duty and following orders in scheduling trains filled with people headed for the extermination camps.

The Nazi architect and administrator Albert Speer claimed that modern technology, not his own morality, was the cause of his behavior. Everyone is just doing his or her job. The complexity of modern technological systems makes it progressively more difficult to assign responsibility for any particular result. Numerous people at various levels and numerous complex devices are responsible for any particular result. When we receive a misstated bank balance the bank teller will usually attribute it to "computer error," but who or what was responsible for the error is rarely determined. With the use of computer programs that learn and evolve on their own even the experts cannot determine the detailed source of some results of the programs.

Criticisms of the Autonomous Technology Thesis

Although Ellul makes many valid particular points, one can raise doubts about a number of Ellul's sweeping conclusions. Ellul's claims about the general lack of competence of any one individual to master both the technical details of an advanced technology and the social, political, and ethical issues that it raises still hold true. But since Ellul wrote his major work in the 1950s there has been a great deal more effort and activity attempting to deal in an informed way with the social issues of technology. Even in the 1950s, whether or not nuclear physicists such as Bohr and Oppenheimer were politically naive, there were other technologists such as Edward Teller who were politically shrewd (although the more politically adept ones were on the side of supporting nuclear technology, not limiting it). In the case of the Human Genome Project there has been a great deal more public discussion of the social and ethical issues than in previous scientific and technological projects. A portion of the funding for the project itself was specifically devoted to its ethical, legal, and social issues (ELSI). Much of the work in this area has involved collaboration of scientists, lawyers, and ethicists, partially overcoming the areas of ignorance that Ellul claimed each group has. Cynics have claimed that this was simply a move by the scientists to attempt to cover themselves and defend against what they realized would be widespread public fears about biotechnology in general and human genetic engineering in particular. Nevertheless, panels of biologists, social scientists, lawyers, and ethicists have been formed, issue reports, and affect public opinion.

We have seen that the view of science as automatically making a predetermined sequence of discoveries, determined solely by the nature of reality and a mechanical scientific method, can be questioned. Scientific theories and foci of interest are often guided by more general climates of social opinion. Interpretation of data is similarly influenced by social context. Similarly, the view of technology as simply a matter of applying science can be questioned. The application of science to technology is not simple and automatic. Many technological discoveries involve an element of chance even when imbedded within a framework of scientific knowledge. The topics to which science is applied are often influenced by the desires and problems of the society. How a technology is used and maintained is also influenced by social context.

It is certainly true that big technological projects supported by powerful political and economic interests can steamroller and intimidate or discredit grassroots opposition and use the media to sway public opinion. Nevertheless, it is not always clear that they succeed. The difficulties and opposition that the nuclear industry has encountered in Germany and the USA (although not in St Simonian technocratic France and Russia) show that cultural issues, attitudes toward nature, and public fears can influence large economic projects that have government support. Similarly, the decline in funding for space exploration after the end of the Cold War shows that the technology alone does not automatically expand once the cultural context has changed (in this case the decline of the nationalistic rivalry between the USA and USSR). Although advertising is powerful in swaying consumers, well funded and advertised products can fail to find buyers. A famous example of this is the Edsel car, which was certainly well advertised by Ford, but which flopped. As for the reverse adaptation thesis, one needs to separate those social forces that are part of a particular technological system from those in the larger society. Often it is not the technological system itself (for instance, NASA in the case of the space program) but the national government that influences the success or decline of the technological project.

Study questions

1 Does Winner's notion of "reverse adaptation" accurately describe the effect of major technological systems on society? That is, do technological systems mold the opinion of politicians, consumers, and others to support and further the aims of the system? Is Winner right about such things as space exploration or military technology? Give an example or counter-example different from those in the text.

2 Does Ellul's claim that technologists are ignorant of politics, while politicians and business people are ignorant of technology, and citizens in

general are ignorant of both the technical and social aspects of technology, still hold true today? Do computer and biotech startup firms run by scientists and technologists refute his claim? Does the growth of popular science and technology writing show that citizens are more informed than Ellul claims?

3 Do the occasional failures of major marketing strategies (such as Ford Motor Company's Edsel car, which was widely advertised, but failed to sell) show that Winner's "reverse adaptation" thesis is wrong?

8

Human Nature:
Tool-making or Language?

A number of characterizations of human nature in philosophy, history, economics, and other fields in recent centuries have focused on tool-making, placing technology as central to human nature. The major alternative characterization of human nature in opposition to this tool-making or technological characterization during the past century has been in terms of language. Humans certainly are both tool- and language-using creatures. Those philosophers of technology who are negative or pessimistic concerning technology are usually those who define humans in terms of language in order to reject the notion that technology is central to being human. Both the tool-making and language views, in order to claim to be essential characterizations of humans, must claim that human tool-making is different from animal tool-making and that human language is qualitatively different from animal communication systems.

In classical philosophy, using real definitions and theories of essences or natures, there were attempts to characterize human nature. Aristotle famously defined humans as rational animals. Some existentialists would reject any such essential definition of humans. Jean-Paul Sartre claims that human freedom precludes any characterization of human nature, as humans may chose to reject any particular characteristic. Simone de Beauvoir has encapsulated this view with the claim that a human is "a being whose essence is not to have an essence." Ortega y Gasset has claimed that humans do not have a nature but have a history.

Nonetheless, many writers have attempted to characterize the human essence. While some center their characterization on tool-making, others have denied that tool-making is central to human nature, instead claiming that mind, language, or symbolism is the most important characteristic of humans.

Usually the side a writer takes on this debate (tools and technology or mind and language) depends on the writers' attitude to technology. Thinkers who see technology as primarily beneficial take the position that humans are primarily tool-makers. Thinkers who seem technology as a danger or a curse tend to emphasize human mind (in earlier writers) or language (in twentieth-century thinkers) to counter the pre-technology stance. Lewis Mumford, Martin Heidegger, and Heidegger's student, Hannah Arendt, emphasize language and art as characterizing human nature or the human condition in order to counter the dominance of technology in an understanding of the nature of humanity.

Head versus Hand: Rationality versus Tool-making as Human Essence

Some of the traditional characterizations of human nature and disputes about such characterizations involve technology. Benjamin Franklin defined or characterized humans as tool-using animals. Others have characterized humans as tool-making animals. The debate about the priority of the mind and a purely mental notion of rationality (recall Aristotle's "Man is rational animal") goes back to the origins of Western thought.

Anaxagoras (c.500–428 BCE) claimed that the development of the prehensile hand preceded the development of the human mind. The later systematizers of ancient philosophy rejected this early idea of the priority of the hand and of physical manipulation to mental contemplation. Aristotle in his *Parts of Animals* (X 687a5) denied Anaxagoras' claim as absurd and thought it obvious that the hand is present to serve the mind. Plato in the *Timaeus* 44d, his account of the creation of the universe, fancifully claimed that humans were first made as heads alone. Part of the reason for this was that heads are circular and the circle is the most perfect figure, the model for the paths of stars and planets. However, these heads got stuck by rolling into ditches and depressions in the ground and so the demiurge creator modified them by adding limbs.

The ancient Greeks and Romans generally denigrated manual work (see box 8.1). Since work was done largely by slaves, manual labor was considered lowly and unworthy of a free man. Plato and Aristotle exalted contemplation and pure theory. Philosophy, mathematics, and astronomy were worthy objects of contemplation as they involved pure theory.

Box 8.1

Ancient, medieval, and early modern attitudes to work and technology, East and West

Work is certainly central to technology. One recent text in the philosophy of technology even goes as far as to define technology as "humanity at work" (Pitt, 2000, p. 11). This approach is understandable, although it omits the use of technological devices for games and sports. The centrality of use of tools and machines in work leads to such a characterization. Likewise, work and the organization of work are central to thinking about technology. However, it is useful to remember that the high valuation of work in the modern Western world was not held universally. Traditional Greek, Roman, and Chinese societies did not revere hard manual work, but looked down upon it.

The tradition of growing very long fingernails in late traditional Chinese society, even in the early twentieth century, was a symbol that the bearer of such impractical fingernails did not have to do manual work (which would be impossible in such conditions). This was despite the amazingly high quality of ancient and medieval Chinese technology, which in many areas was far ahead of that of Europe and in some cases (such as hormone therapy and biological pest control) was even ahead of Europe until the twentieth century (Needham, 1954). (See the section on China in chapter 10 for more on this topic.) Despite the innovations of medieval Chinese technology, the craftspeople responsible for this technology were looked down upon throughout the Chinese Empire (from approximately 200 BCE). No school of Chinese philosophy except for the Mohists revered technology. They are thought to have been military engineers, and in some cases even slaves, who flourished during the period of contending schools before the rise of the Empire imposed an official philosophy. Even the early Taoists, who were responsible for chemical and medical advances, in the works attributed to Lao Tzu and Chuang Tzu spoke ill of labor saving devices.

In ancient Greece and Rome, a similar situation prevailed from the classical age of Greece around 400 BCE until the late Roman Empire around 400 CE. Before the rise of the massive use of slavery the early, pre-Socratic Greek philosopher-scientists (so-called because they pre-dated Socrates) used metaphors from technology in a positive way in their metaphysics. For instance, Empedocles is said to have eliminated disease in his city through draining the swamps by diverting a river. However, by the time of Plato and Aristotle and throughout the rest of the period of the Greeks and Romans, manual labor was looked upon unfavorably. For Plato and Aristotle the contemplative life was the highest form of life (see the discussion of Plato in chapter 3). Reflection on philosophy, astronomy, and mathematics was the highest calling. Manual work was fit only for slaves, and its association with slavery degraded its status. According to one anecdote, Plato expelled two members of the Academy because they had invented a curve-tracing device. Plato thought this mechanical way of doing geometry was vulgar. The student should study the forms purely by mental contemplation.

Plutarch in his *Lives* famously says of Archimedes that he focused on abstract mathematical theory and distained to record his practical inventions, which were the diversions of a mind at play. Despite the fact that Archimedes developed numerous inventions, including most notably military devices such as catapults, burning mirrors, and other weapons in an attempt to stave off the siege of his home city Syracuse, he never bothered to record these inventions (Farrington, 1964, p. 216). Furthermore, later Greeks, such as Plutarch, cite with approval Archimedes' reluctance to honor such technological devices by bothering to record them as he did his more abstract mathematics and mathematical physics. Several anecdotes, although somewhat garbled, note that Roman emperors refused to reward labor-saving inventions and even punished some inventors with death (Finley, 1983a, pp. 189, 192). Likewise, although the Greeks were careful to record the inventors (real or mythical) of literary devices and concepts, they hardly ever recorded the inventors of even the most important technological devices (except for Prometheus' mythical gift of fire to humans).

With the decline of slavery and the dominance of Christianity in the Middle Ages, work took on a more positive value. Although the story of the expulsion of Adam and Eve from the Garden of Eden in the Book of Genesis (3: 16–17) in the Bible presents work "by the sweat of thy brow" (along with Eve's future labor in childbirth) as a punishment, medieval society raised the status of work relative to that of traditional Greek, Roman, and Chinese society.

The monastic Order of Saint Benedict, with its Rule, was one source of the expression of the higher status of work (Boulding, 1968). The monks in the medieval monasteries had to support themselves and do their own work. The monks developed a number of technological devices, including pulleys (White, 1962). They allegedly even developed an alarm clock to wake up monks for middle of the night prayers (Mumford, 1934, pp. 12–18; but see also Landes, 1983). Whether by coincidence or world system dynamics, Zen Buddhist monasteries in medieval China, thousands of miles away from Europe, during the same decades of the late eighth century, developed a similar Rule of Paijing, emphasizing work and monastic self-sufficiency, which helped to gain sympathy and support from the surrounding peasants during government suppression of Buddhism.

Work received higher value with the rise of Protestantism, particularly Calvinism. Sociologist Max Weber (1864–1920), in his *The Protestant Ethic and the Spirit of Capitalism* (1904), said that the notion that people were predestined from creation to be either saved or damned created great anxiety. It was thought that wealth and success in this world was an outward sign that one was saved. Certainly the subjectivity and emphasis on conscience of Protestantism also contributed to the individualism of capitalism. Weber notes at the end of this book that although religious motivation declined, the structures of capitalism remained, rather like a snail that secreted a shell, and then the religious animal died, leaving the shell, which, Weber suggested, might become our "iron cage" (Weber, 1904, p. 181).

Though these early views may seem primitive or even to many moderns comical (in the case of Plato on human evolution), they foreshadow the continuing debate about priority of theoretical thought or of practical technological action in the characterization of human nature.

Haeckel, Engels, and Paleoanthropologists on Humans as Tool-makers

The debate concerning the priority of head and hand was renewed in the late nineteenth and early twentieth centuries in anthropology. The German evolutionist Ernst Haeckel (1868, 1869) claimed that the upright stance freed the hands for manipulating the environment and this led to the evolution of increased brain size. In "The part played by labor in the transition from ape to man," Friedrich Engels, Marx's collaborator, followed Haeckel in this claim. Engels further argued that the emphasis on the priority of the mind was a product of the higher evaluation of mental over physical labor in class-divided societies (Engels, 1882). The priests and administrators engaged in mental labor and considered themselves superior to the peasants and crafts workers. Engels tied the evolutionary priority of the hand over the brain to a general theory of history in which manual labor is the basis of human society and in which the development of technology, along with class relations, drives human history.

The debate was, basically: did humans first get smart and then stand up, free their hands, and make tools? Or did they first stand up and make tools and then get smart?

At the turn of the twentieth century many British anthropologists, such as Wood Jones and Grafton Elliott Smith, claimed that large brains came before hands in human evolution (Landau 1991). They implicitly allied themselves with the two millennia long classical notion of the priority of the human mind over the hands and body that we saw originally formulated in Plato and Aristotle. This theoretical bias contributed to the acceptance of the fraud of Piltdown Man. Piltdown Man was a planted fossil consisting of the cranium of a modern human attached to the jaw of an orangutan. This supposed fossil gave credence to the belief that the earliest humans had large brains but an ape-like body that later evolved the characteristics of the human body. Because it supported their pet theory, a number of British anthropologists in the early twentieth century were insufficiently critical of Piltdown. Although Piltdown was "discovered" in 1913 and succeeding years, it was not exposed as a hoax until the early 1950s (Weiner, 1955).

Twentieth-century anthropology has lent support to the thesis that, in Sherwood Washburn's words, "Large brains followed tools." Technology drove encephalization. Nevertheless, debate has continued as to whether brain size increase led to tool-making or whether tool-making led to brain size increase. In 2003, stone tools along with animal bones showing cuts from stone tools were found that date to 2.6 million years ago (*New York Times*, October 21, 2003, p. D3). This supports the view that tool-making preceded brain size increase, but since the most likely pre-humans present at this site were Australopithecines, it suggests that the tool-making definition of humans would include proto-humans prior to the genus *Homo*.

The definition of humans as tool-makers has influenced the interpretation of evidence for the emergence of humans and guided paleoanthropologists (those who study of human fossils) in their identification of the line between humans and non-humans. When Louis Leaky discovered very early stone tools, before even finding the skeletons of their makers, he designated the maker as *Homo habilis*, or handy man. Leaky simply *assumed* that tool-making characterized humans and that the first tool-maker was the first human.

Animal Technology as a Counter to the Uniqueness of Humans as Tool-makers

Certainly humans have been involved in technology from the very beginning. However, in the twentieth century, discovery of animal tool use has been used against the claim that humans are *the* tool-using animals. *Sphenx* wasps use a pebble to tamp down the soil on the hole in which they bury their prey. Crabs use sponges for camouflage. One of Darwin's finches uses a blade of grass to poke into holes to draw out insects. Chimpanzees use twigs to poke into holes in trees to extract termites to be eaten. Chimpanzees also break off twigs for catching insects and break off branches for combat. Thus chimpanzees can also be considered tool-makers, if of only the most rudimentary sort.

Lewis Mumford has noted that the overemphasis on hand tools and the de-emphasis on container technology in the study of human society has led to a minimalization of animal tool use and tool-making. The nests of birds and of paper-making and mud-daubing wasps, the dams and lodges of beavers, are treated as tools or products of tool-making.

Nonetheless, human tool-making has a characteristic that makes it different from animal tool-making. Humans make tools to make tools. Human language differs from animal language in its grammatical recursiveness, its ability to be indefinitely extended by further substitutions. Human tool-making also partakes of this recursiveness. There are tools that make tools that make tools . . . It has even been suggested by Patricia Greenfield (1991) that the part of the brain of apes used for digital manipulation by the hand was taken over, in part, for language in the evolution of humans.

The Criticism of Man as Tool-maker in Twentieth-century Philosophy of Technology

A number of twentieth-century writers on technology who are critical or skeptical of the promise of technology have championed the claim that language, not tool-making, is what is characteristic of humans. Language, as the realm of meaning, is held up in opposition to technology. The claim about language as characteristic of humans is often part of a strategy to de-emphasize the centrality of technology to what is valuable about humans. A number of these writers are generally thought to hold negative views of technology. However, some, such as Lewis Mumford, favor an alternative, decentralized technology to the dominant one, which they consider to en-courage centralization, top-down rule, and anti-democratic tendencies.

Karl Marx on Human Nature, Technology, and Alienation

Karl Marx in his early writings presented a theory of human nature centered on creative labor. In his later writings Marx presented a theory of society and history centered on the role of labor in the reproduction of society. Because of Marx's emphasis on the role of technology as the basis of the succession of forms of society throughout history, many philosophers of technology have grappled with Marx's ideas. Anti-Marxists as well as Marx-ists have criticized his politics while borrowing one or another of his ideas. One historian has claimed that his colleagues denounced Marx in the class-room by day and ransacked his works for ideas for their own research by night (Williams, 1964). Much of later European philosophy of technology, like classical sociology, can be called "a debate with Marx's ghost" (Zeitlin, 1968). Just as Whitehead called Western philosophy a series of footnotes to Plato, so Robert Heilbroner (whom we met in chapter 6) has claimed, in effect, that much of twentieth-century social science consisted largely of footnotes to Marx (Heilbroner, 1978).

Joseph Schumpeter, the conservative Austrian-American economist, wrote of Marx's influence, even on those, like himself, who rejected his conclusions:

> Most of the creations of the intellect or fancy pass away for good after a time that varies between an after-dinner hour and a generation. Some, however, do not. They suffer eclipses but they come back again, and they come back not as unrecognizable elements of a cultural inheritance, but in their indi-vidual garb and with their personal scars which people may see and touch. These we may well call the great ones – it is no disadvantage of this definition that it links greatness to vitality. (Schumpeter, 1950, p. 3)

The view of humans as tool-makers that we find in Engels and in much of later Marxism is *not* Marx's own view of human nature. Marx explicitly

119

presents his view of human nature mainly in his *Early Writings* (1963). Technology is central to Marx's account of economics and human history, but Marx, unlike Engels, never claimed that humans are primarily tool-making animals. Marx's complex, ambiguous, and possibly changing doctrine of human nature in relation to technology reflects his belief that technology is the key to human prosperity and liberation from drudgery in future communism, even though at present technology is the means of exploitation and oppression of workers under capitalism.

Although Karl Marx has been commonly described in the socialist and communist movements as denying the existence of human nature, writers influenced by his *Early Economic and Philosophical Manuscripts* (or *1844 Manuscripts*) have described Marx's theory of human nature. Part of the disagreement about this theory is due to the fact that Marx's published works, which were for the most part his later works, often disparaged theories of human nature held by traditional philosophers and by Marx's economist contemporaries. He ridiculed Benjamin Franklin's and British historian and cultural critic Thomas Carlyle's (1795–1881) characterization of man as a tool-using animal, along with Bentham's account of humans as solely driven by pleasure and pain, and what was later to be called the "economic man" model of mainstream economists.

In the later works that were published during his lifetime, such as *Contribution to a Critique of Political Economy* (1859) and volume 1 of *Capital* (1867), Marx has few or no explicit comments about human nature. Marx, as well as the later leftist tradition, has tended to emphasize the cultural and social variability and mutability of human characteristics, institutions, and behavior. "Orthodox" Marxists in the social democratic parties of the nineteenth century and the communist parties of the twentieth century claimed that Marx denied the existence of human nature. Marxists, as well as utopian radicals, claimed that people under communism would lack many of the selfish and greedy characteristics that conservatives, traditional economists, and cynical "common sense" attribute to us. Because of this Marx's followers tended to deny and to claim that Marx denied any human nature. Part of the reason for this is that the common-sense phrase "that's human nature" tends to be uttered with a shrug about undesirable characteristics of human behavior and reinforces a kind of unreflective conservatism. Opponents of Marxism, whether traditional religious writers, economic defenders of capitalism, or defenders of biological theories of human nature (such as social Darwinism or sociobiology), often agreed with orthodox Marxists in denying that Marx had a doctrine of human nature. In the second half of the twentieth century

HUMAN NATURE: TOOL-MAKING OR LANGUAGE?

Marx's biological determinist opponents have often claimed that his views espouse an environmental determinism like that of Skinner (Wilson, 1978) (see the early part of chapter 6 on determinisms).

The discovery and publication of Marx's *1844 Manuscripts* in 1931 (in the original German) and around 1960 (in English) changed many people's conception of what Marx had said about human nature. In these early works, unknown for almost ninety years, Marx speaks of human "species being," an obscure phrase that certainly seems to connote a human nature. Marx also goes on at length about human needs and human characteristics and the way that they are deformed and distorted by "alienation" under capitalism. All this seems to support the view that Marx held a theory of human nature as something that could be alienated. The psychologist and member of the Frankfurt School Eric Fromm, when he published these works in English, called the collection *Marx's Concept of Man* (Fromm, 1961) and saw them as contributing to a "Marxist humanism." One of the first reviewers of Marx's early manuscripts when they first appeared in Germany was Herbert Marcuse, who was highly influenced by them (Marcuse, 1932) (see the discussion of Marcuse and the Frankfurt School in chapter 4).

Nevertheless, Marx's views on human nature are complex and ambiguous (Marx's critics would say they are inconsistent and confused). Marx speaks of human needs, but also writes (in a later, unpublished, draft of *Capital*) that, as society develops, humans develop new needs. Human nature or "species being," whatever it is, does not seem to be an essence located or instantiated within each individual, as many traditional Aristotelian and rationalist views imply. In fact, in his "Theses on Feuerbach" (a fragment slightly later than the *1844 Manuscripts* but earlier than the books published during his lifetime) Marx states that the human essence does not reside within each individual, but that "man is an ensemble of social relations." Many traditional Marxists claim that this shows Marx's early break with his "juvenile" humanism (Althusser, 1966). (The German Democratic Republic's Communist Party edition of the complete works of Marx placed the early writings in an appended volume of "*juvenilia*," whose publication was delayed for decades.)

However, an emphasis on relations does not necessarily mean the denial of essence or nature. Traditional essentialism centered on properties or qualities of the individual object. Relations were treated as secondary to and derivative from qualities in most traditional philosophy. With Hegel, consequently with Marx, even more so in American pragmatism and process philosophy, and independently in modern symbolic logic, relations came to be treated as equal to, or even prior to, qualities of objects (see the discussion of "process philosophy" in chapter 12). Marx's "species being" may itself be a relational complex, rather like the so-called biological species concept, defined by evolutionary taxonomist Ernst Mayr (1905–2005) as a potentially mutually interbreeding population (Mayr, 1957), without denying that one can characterize the human species.

Against the common claim that Marx thought human nature was so infinitely malleable or manipulable, one can ask: why did Marx think that something was wrong with capitalism? Indeed, the very concept of alienation involves the notion of a true self or nature from which one is alienated. Kinds of alienation that Marx discusses include alienation from the product of labor (lack of ownership and control over the product), alienation from the labor process (lack of control over the conditions of work), alienation from one's own species being (including alienation both from oneself and from external nature), and alienation from fellow humans. Some of Marx's characterizations of human beings seem to present a goal-oriented or teleological account of human nature. Human nature is the characterization of humans as they will essentially be in the future, when they have overcome alienation. Nevertheless, some of the accounts Marx gives of human nature emphasize attributes that are at least partially present in humans now. Marx emphasizes that humans are active, praxis-oriented beings that transform their environment. This activity is primarily manifest through human labor, the means by which humans transform their world. This labor, when unalienated, is creative and fulfilling. Marx in his later works, such as volume 3 of *Capital*, published after his death by Engels, seems to partially retreat on this. He ridicules the utopian socialists for thinking that work can be turned into play, and he claims that there will always be a residue of necessary labor, which remains unfree. Hannah Arendt criticizes Marx's alleged multiple ambiguities about labor (see the section on Arendt below).

One major respect in which Marx differs from many later writers, such as Ellul, Mumford, Heidegger, and Arendt, is that he does not see physical technology as *necessarily* leading to alienation and the problems of the modern world. For Marx, capitalism is the problem. Marx is a theorist of capitalism, not, as were St Simon, Comte, and the technocrats, a theorist of industrial society as such. The types of technology developed by capitalism and the uses of technology for labor discipline under capitalism are the major problem according to Marx.

However, there are further ambiguities with Marx's view concerning the respective responsibilities of technology and of capitalism for exploitation and alienation. In *Capital*, Marx distinguishes between the technological division of labor and the social division of labor. The former is the division of work tasks imposed by the nature of the machines used in the factory. The latter is the division of labor imposed by capitalists for the purpose of controlling the workers by centralizing planning, decision-making, and knowledge of the overall production system in the foreman's office. Thus workers lack the knowledge and control of the production process that would otherwise allow them to think they could run it themselves. Much late twentieth-century Marxist analysis of the labor process, such as that of Harry Braverman (1974) and Steve Marglin (1974), centers on this transfer of skills and knowledge from the workers to the managers, what British writers called "deskilling." However, drawing the line between those aspects of the division of labor due purely to the needs of production and those due to the need for control of workers is difficult.

Engels late in life, writing against the anarchists in the article "On authority" (Engels, 1874), argued that technology alone makes labor discipline necessary. Indeed, technology as such produces "a veritable despotism independent of all social organization." In reacting to the anarchists' demand for autonomy of the workers, Engels gives up all consideration of alternative technologies and alternative forms of factory organization that would be less authoritarian than the present ones. Lenin makes much of Engels's article to justify factory discipline in the USSR. Ironically, Lenin praised the American time and motion studies advocate Frederick Taylor, and even claimed that Taylorism was central to socialism. Thus, ironically, Lenin, seeing "society as one vast factory," using St Simon's phrase, identified the most mind-numbing control of the minute motions of the laborer with socialist liberation.

Late twentieth-century Marxism, more in the tradition of Braverman and Marglin, above, has envisaged the possibility of different forms of factory organization and different forms of technology that would make labor control less authoritarian. Braverman notes the use of computer surveillance of various sorts for the monitoring, disciplining, and speeding up of white-collar work.

Lewis Mumford against the Characterization of Humans Primarily as Tool-makers

Some thinkers look at contemporary technology primarily as a bane rather than a blessing. The American Lewis Mumford, the German philosopher Martin Heidegger, and his student Hannah Arendt all counter the conception of humans as tool-makers, laborers, and technological animals with the characterization of humans as linguistic and symbol-creating creatures.

Lewis Mumford was a prolific writer on American architecture and world cities, as well as on technology in general. An independent intellectual who never held a permanent academic position, Mumford undertook interdisciplinary work combining the history of technology, philosophy, literature and art criticism, anthropology, and sociology. He moved from a more positive evaluation of contemporary technology in the 1930s (Mumford, 1934) to a much more negative one in the 1960s (Mumford, 1967), influenced in part by the nuclear standoff and threats of nuclear war in the Cold War. Although critics have called him anti-technology, he claimed to be advocating a more decentralized technology of human scale. He calls the dominant form of technology authoritarian technics, tied to centralization, top-down organization, and oppression, as opposed to an alternative that he calls democratic technics, which is liberating and life-affirming. Mumford was a follower of the Scottish biologist Patrick Geddes, who took an ecological approach to the study of cities at the beginning of the twentieth century. They advocated regional planning to integrate the city with its rural surroundings, and garden cities (mixing green spaces with urban buildings). Mumford's ideas were precursors of and have many affinities with the ecology movement, bioregionalism, and the small technology movement.

Mumford imagines human society prior to the rise of the state as life-affirming and egalitarian. Females participated centrally in the gathering and storage of food. He sees the move from big game hunting to the more settled Neolithic existence as leading to a desirable condition. Technologies devised by women, such as baskets and pottery, were central to supporting human communities. Womb-like containers and caves were as important as phallic spears and arrows. Tools were used but did not dominate other aspects of human life. The human body, its role in dance and ceremony, and as an object of body painting, piercing, and decoration, was a central means of creativity. Mumford claims that the most important "tool" of early humans was the human body.

Mumford sees the aesthetic and metaphorical aspects of human thought as the primary source of early human advance. In contrast to a picture of increasing abstraction and rationality as the key to human advance, Mumford suggests that dream-life, with its often wild associations of images and ideas, despite its dangers, was the source of early innovation. He thinks that the Australian aborigines' "dreamtime" is representative of the source of cultural advance in early humans.

HUMAN NATURE: TOOL-MAKING OR LANGUAGE?

Mumford sees the overemphasis on the tool for understanding human nature as contributing to a vision of technology-dominated humanity. Mumford himself claims that the first "machines" were made not of wood or metal but of large numbers of human bodies organized for work. The earliest civilizations engaged in huge construction or irrigation products. Egyptian pyramids, the Great Wall of China, the canals of the Tigris–Euphrates, and the dams for flood control in China are examples of gigantic construction efforts in the earliest civilizations. A God-King ruled these first civilizations, a king identified with the gods, and a bureaucracy of priests possessing secret knowledge, as in the case of the astronomer-priests of ancient Egypt, China, and the Tigris–Euphrates valley (what is now Iraq). Many of these states, such as the early Egyptians and the Assyrians, engaged in brutal conquest and carved monuments to the murderous power of their divine rulers. Mumford calls the first machine the "megamachine" (see the discussion of technology without tools in chapter 2).

Mumford claims that twentieth-century society, with its totalitarian states with absolute dictators and, even in the democratic nations, welfare bureaucracies and huge military–industrial complexes with secret knowledge of scientists and national security experts (instead of priests), more resembles the first civilizations than it does the more decentralized, competitive, market societies of the nineteenth century. He compares the death-oriented calculations of modern experts on nuclear weapons and warfare to the sadistic boasts of the ancient empires' God-King conquerors in their threats to competing states.

MIT nuclear engineer Alvin Weinberg writes that there is a need for a "nuclear priesthood" and a need to build huge monuments (Weinberg suggested pyramids as a possible model) over nuclear waste storage areas to warn future generations (Weinberg, 1966). This certainly fits Mumford's vision of the twentieth-century USA as a modern version of the Egypt of the pharaohs. Mumford also compares the Cold War nuclear superpowers, the USA and USSR, to the aggressive and secretive military empires, such as that of ancient Assyria, in what is now Iraq. Mumford titles the second part of his last magnum opus "The Pentagon of Power," playing on the role of the pentagon in witchcraft and the name of the administrative center of the US military. For Mumford, talk of scientists as the "new priesthood" (Lapp, 1965) and nuclear strategists as the "wizards of Armageddon" (Kaplan, 1983) is more than empty metaphors in book titles. Auguste Comte, with his projected technocracy of priestly scientists, looked forward optimistically to what Mumford sees as a negative, death-oriented military–industrial complex with its corps of scientific advisors and spokespeople (see the discussion of Comte in chapter 3).

In a positive vein, Mumford wishes to return to something closer to the life-affirming technology of our Neolithic hunter-gatherer ancestors. He envisages a more decentralized technology, "democratic technics" replacing the centralized and bureaucratic technology of today. He thinks that demoting the tool-centered conception of humanity and promoting understanding humans as symbolic animals is a theoretical contribution to this transformation. Human beings for Mumford are not primarily tool-making animals but are symbolic animals.

Martin Heidegger

The philosopher **Martin Heidegger** (1889–1976) was probably the leading and was certainly the most influential German philosopher of the twentieth century. Heidegger was early struck by the problem of understanding Being in Aristotle. He combined a concern for the classical problem of being in the earliest Greek philosophers with an interest in the insights into the human experience of anxiety and the human condition of nineteenth-century existentialist philosophers such as Søren Kierkegaard and Friedrich Nietzsche. Heidegger absorbed and considerably modified the phenomenological method of his teacher Edmund Husserl with existential notions of human life and with hermeneutics or interpretation of meaning (see chapter 5). His philosophical approach led him to reject traditional attempts to define a human essence in the manner of Aristotle or to describe humans primarily in terms of mind or mental substance, as did the early modern philosophers, but he did attempt to characterize the human condition.

Heidegger contrasts language as the realm of meaning, "the house of Being," with the technological realm. For Heidegger technology is the primary characteristic of our age, replacing the notions of biological growth, artistic creation, or divine creation that characterized previous epochs. The development of Western philosophy from the time of Plato, at least, has been a trajectory toward the domination of the world by modern technology.

Despite the centrality of technology to the present epoch and the destiny of Western Being, humans are not fundamentally tool-making animals according to Heidegger. Furthermore, humans are not "rational animals." Indeed, humans are not animals, and an account of human being fails if it is based on the evolution of humans and their rationality from animals. For

HUMAN NATURE: TOOL-MAKING OR LANGUAGE?

Heidegger, animals lack the world-openness and world-constructing character of humans. Since they are devoid of language in the full sense they are incapable of understanding the world and things. Heidegger rejects Aristotle's definition of humans as rational animals, but his account of the role of the human hand in relation to consciousness interestingly resembles Aristotle's. Just as Aristotle rejected Anaxagoras' claim that the mind evolved because of the development of the hand and claimed that the hands exist for the sake of the mind (see above), Heidegger claims that the human hand discloses things in a way that the animal paw does not. Work, tool use, and tool-making only occur within the context of language, which animals lack. Tools and even human hands are not "organs" of humans the way that McLuhan and many others claim. Tools, in themselves, are ahistorical and human hands are disclosive; that is, they reveal a human world.

Originally, some suggest, Heidegger saw the solution to the technological age in an aesthetic relation to technology, in which technology would become one with art (Zimmerman, 1990). Later he seems to have seen the solution in replacing our "enchained" relation to technology with our achieving a "free relation" to technology. What exactly this free relation is has been a matter of dispute among followers of and commentators on Heidegger.

In Heidegger's earlier thought the meaningful experiences of traditional peasant culture (in craft work and agriculture) are contrasted with the modern technological attitude. Albert Borgmann, a present-day American philosopher of technology, suggests that Heidegger's rural and archaic references can be replaced with "focal experiences" and communal celebrations in contemporary life, such as preparing a meal or jogging. In Borgmann's view these meaningful focal experiences should be central to our values and only served by technological means (Borgmann, 1984).

Dreyfus (in Dreyfus and Spinosa, 1997) suggests that new attitudes and orientations that do not reject technology but also embrace non-technological, traditional, or communal practices and attitudes become possible after one recognizes that we are enmeshed in the technological understanding of being. This understanding, or "releasement," is not sufficient, but is a necessary precondition for the embracing of other attitudes. Dreyfus uses the example of Japan, where embrace of technology coexists with traditional cultural practices. This apt example of Japan is also used by Feenberg (1995, chapters 8 and 9). Dreyfus also mentions Woodstock (which, admittedly, failed in its highest ambitions to transform society), where the technology of music amplification was put at the service of fostering a community. Dreyfus, in this article, also endorses Heidegger's claim of the need to replace calculative with meditative thinking.

Dreyfus and Spinosa (1997) move even further away from Borgmann than in Dreyfus's earlier article when they characterize the "free relation" to technology in terms of the decentering or pluralization of the self caused by technology. They present a postmodern version of Heidegger's "free relation," claiming that we can live in a plurality of local "worlds" provided by technology (see box 6.3). They use examples from Sherry Turkle's *Life on the Screen* (1995), describing the multiple identities serially assumed by users of chat rooms on the Internet. They also refer to the temporary joining and creative reinforcement of members of rock groups and technological research groups and the subsequent separation of members to join other groups.

Andrew Feenberg, the critical theorist, follower of the Frankfurt School (see Habermas, below, and chapter 4), criticizes both Borgman and Dreyfus and Spinosa as excessively idealist in their solution to the problem of living with technology. They understand the solution to reside in a new attitude, and in the supplementing of technological existence with other communal forms of existence, but they do not ever propose redesigning technologies themselves. The environmental, military, and labor discipline effects of technology would remain, only to be supplemented by intimate or communal acts that would allow an alternative perspective. (To be fair to Dreyfus, in his earlier article he does say that work on environmental harms of technologies should go on, but is less important than the change of fundamental attitude.)

After Heidegger, interpreters like Borgmann have emphasized the non-technological experiences of meaning that help to allow us not to be unreflectively engulfed in technology, while others, such as Dreyfus, have emphasized the ability of postmodern selves to free themselves from any predetermined characterization by making use of the resources of technology. It would seem that Heidegger's own views reside somewhere between the rather essentialistic and emphatically non-technological, ecstatic, focal epiphanies of Borgmann, and the totally unmoored and rootless self-transformations or morphings of Dreyfus.

Hannah Arendt on Work, Labor, and Action

Hannah Arendt, a German-American, was a student and lover of Martin Heidegger. A Jew, she fled Nazi Germany, and eventually settled in the USA. Her work on modern society is highly influenced by Heidegger and his vision of the dominance of technology in the modern era. Arendt, perhaps even more so than Heidegger and Mumford, is totally negative about the prospects of modern technology. Heidegger hints at a new attitude to technology that will allow us to use technology while freeing us from domination by it. Mumford suggests an alternative, utopian, decentralized biotechnics and democratic technology. However, Arendt offers no such hope for future improvements of technological society.

Arendt called her 1958 book *The Human Condition*. She discusses the human condition rather than human nature, in part because of the influence of Heidegger. Like Heidegger, and the existentialists noted at the beginning of this chapter, she rejects the Aristotelian notion of an innate and unchanging human essence. She also believes that human nature has to be a theologically based concept, and hence she rejects the concept, as she doesn't ground her views in religion. Arendt's first book was *Augustine's Doctrine of Love* (1929). However, like Heidegger, following Nietzsche, she thinks Judeo-Christianity is *passé*.

Arendt contrasts what she calls **labor, work**, and **action**. In the background of this distinction is what she claims is ambiguity in Marx's notion of labor. On the one hand, for Marx, unalienated labor is creative and positive. Labor is what produces the human world and environment. Labor is for Marx the "Heraceitian fire" of human personal creative energy. On the other hand, labor is burdensome and exhausting. Labor as described in *Capital* is exploitative and dehumanizing. In the *German Ideology* Marx even ends the major section by proposing the "abolition of labor." Arendt claims that Marx has *two* kinds of labor that are very different from one another. Furthermore, Arendt claims, Marx tends to conflate political, revolutionary activity and creative *praxis* with labor as creative activity, when labor and political action are very different sorts of activities.

Arendt notes that Greek, Latin, French, German, and English have two words, work and labor, that are generally used interchangeably. Arendt points out that both work and labor *can* denote processes. However, "*a work*" (for instance, "*a work of art*") is a physical *object*, while "*a labor*" (such as one of the labors of Hercules) is an *activity*. Arendt takes this difference to use labor to designate a different activity from work. Labor is pure process, while work culminates in an artifact.

In Arendt's usage, **labor** also does not produce a permanent product, but is the daily repeated activity of life maintenance, primarily household maintenance, including food preparation, cleaning, and bathing. Labor is a biological function in which both humans and animals, as in the search for food, must engage. Unlike action it does not institute something once and for all, but must be repeated continually. In ancient Greece women and slaves performed labor in the household. Arendt makes much of the contrast of public and private in ancient Greece. Women in ancient Greece (with the exception of geisha-like *heterai*, entertainers and prostitutes) were confined to the household and did not participate in the public realm. Free (non-slave) men spent their time in the public arenas of the market place, the law court, and the assembly. Arendt was not a feminist, but feminists after Arendt have made much use of the public man–private woman dichotomy, usually without crediting Arendt and referring to other, more recent, explicitly feminist theorists, some of whom probably borrowed the account from Arendt.

Arendt also plays on the use of "labor" to designate childbirth. In the Book of Genesis (3:16, 19) in the Bible, when God expels Adam and Eve from the Garden of Eden for eating of the tree of knowledge of good and evil, He condemns Eve to the pain of childbirth, parallel to the pain of manual labor for Adam. (One inconsistency of Arendt's here is that childbirth labor *does* create something absolutely new, a new human. However, it is true that reproduction is something humans share with all animals.)

Arendt notes that the Latin term *animal laborans* (laboring animal) identifies labor with an activity not fully human. In contrast, **work**, unlike labor, is a distinctively human pursuit. Work produces objects and constructs the world. While the products of labor disappear upon consumption (as in the case of food), the products of work endure. The objects produced by work are objective and public. Work is the activity of making or fabricating things, as in the crafts. Work is the particular province of free craftspeople (though slaves were also employed in the crafts). According to Arendt, work as fabrication, the production of enduring material objects, existed in the ancient and early modern world, but has largely disappeared in the twentieth century. Artists and individual craftspeople still pursue fabrication in the advanced industrial societies, but this is often a luxury production for yuppies. The vast majority of both assembly line workers in factories and data entry workers in offices no longer work or fabricate in this sense.

For Arendt, **action** is primarily characterized by speech in the public or political realm. Like labor, it produces no permanent physical product but is truly characteristic of the highest form of human activity. In the world before tape recorders and cameras even the most eloquent speech disappeared into the air immediately after its utterance. In contrast to labor, however, action is truly creative. Speech and political action create and institute truly new things, new policies and institutions. At the same time the political and speech acts that create these novelties are evanescent, impermanent.

In Arendt's terminology Marx mistakenly reduced both work and action to labor. Arendt claims that in the modern world action in the true sense has all but disappeared. Genuine participatory discussion in the *agora* (market place) and assembly hall has been replaced by political propaganda. Recall that St Simon (see chapter 6) called artists "engineers of the spirit," which was echoed by Soviet dictator Stalin. In addition, Ellul wrote a whole book called *Propaganda* (1962), and treats propaganda as a form of technology, not communication, which dominates politics and the mass media (see discussion of Ellul in chapters 2 and 7). Likewise, politics in the ancient Greek sense of the term – that is, active, direct democratic participation by the public – has also almost entirely disappeared. Direct participation was appropriate for small city-states, but large states have replaced participation with representation at best. Mass media and bureaucracy undermine even this less direct democracy. In the modern world only briefly in the initial phase of political revolutions does participatory democracy and genuine action emerge. Workers councils in Western European revolutions and the original, briefly flourishing, soviets in the Soviet Union involved genuine participatory democracy and action in Arendt's sense. However, in Western Europe the forces of the old regimes crushed the workers' councils, while in the Soviet Union they were subsumed under the totalitarian state with controlled labor unions and totally lost their participatory nature.

Humans as active speaking animals have been almost completely replaced by humans as laboring animals. Likewise, according to Arendt, work has greatly shrunk in extent in the modern world. Work has turned into labor, as the skilled activity of the individual craftsman has been replaced by the repetitive production of parts of a product on the assembly line in the modern factory. The factory worker producing a single part or making a single adjustment over and over again often has no understanding of the structure of the overall product and cannot take pride in creation (the classic comic portrayal of this is the opening of Charlie Chaplin's silent movie *Modern Times* (1936), in which Chaplin becomes deranged by working on an accelerating assembly line). Planned obsolescence means that the products of work become objects of consumption in the sense of being disposed of shortly after use. Thus the assembly line, with interchangeable parts and factory work yielding disposable items, is more like Arendt's labor (repetitive, with no visible product) than like the craft production of objects identifiable as personal products by artisans.

Arendt's bleak view of the human prospect projects Marx's "metabolization with nature" as a regression to an animal-like, pre-human condition. Aristotle had once joked that if looms could weave themselves, we should have no need for slaves. He didn't envisage modern machinery and power looms. But Arendt claimed, in effect, that rather than technology yielding "Athens without slaves" through the education and leisure permitted by automation, we are left with "slaves without Athens," a world where everyone is a laborer in her sense, via slavery to the machine, and leisure degenerates into mass culture, lacking the high culture of Athens (Robins and Webster, 1990).

Habermas on the Relation of Work to Human Nature

Another German philosopher, Jürgen Habermas, contrasts communicative action with instrumental action. Habermas is of the second generation of the Frankfurt School of Critical Theory (see chapter 4). Habermas began by criticizing and attempting to reconstruct Marx's views, with considerable revisions. Habermas rejects the primacy of work and technology, which he (not wholly accurately) ascribes to Marx, and which was present in most "orthodox" Soviet Marxism. He claims that Marxism neglects the autonomous role of the communication of meanings. Habermas ends up holding a kind of dualism or dichotomy of technological labor and communicative action. In his later works, he does not explicitly present a traditional theory of human nature, but does in his earlier published work write of "species-wide capacities" and similar notions. His unpublished dissertation on Schelling grapples with an evolutionary theory of humans, but does not engage with modern Darwinian natural selection. However, his recent writings on biotechnology assume the truth of a simplistic genetic determinism counterpoised to an anti-scientific "old humanism" (just as he assumes a Marxist-Leninist, mechanistic conception of labor in Marx in order to supplement it with his communication theory). Lenny Moss, a philosopher of biology with a degree in biochemistry, has recently criticized and attempted to revitalize Habermas's early evolutionary anthropology in terms of modern, non-reductive theories of the relation of genetics and the environment (Moss, 2004a, b), using the criticism of reductionistic genetic determinism from Moss's *What Genes Can't Do* (a title borrowed from Dreyfus's *What Computers Can't Do*, discussed in chapter 5).

Like Heidegger, Mumford, and Arendt, Habermas rejects the primacy of technology, but he thinks that technology is of equal importance to symbolic communication in the description of human life. Habermas sees similarities in his communicative aspect of human nature to Arendt's notion of action.

In his *Knowledge and Human Interests* (1970), Habermas presents his theory of "knowledge-constitutive" or knowledge-guiding interests. These "interests of reason" lie behind and motivate the different forms and functions of reason. They are the interest in manipulation and control of nature, the interest in communication and understanding of meaning, and the interest in freedom (the emancipatory interest). Work and interaction are two central activities tied to specific forms of knowledge. Habermas identifies positivist philosophy with a position that implicitly makes the technical-manipulative interest the sole interest of reason (see chapter 3 on the early positivism of Comte and chapter 1 on logical positivism). Habermas identifies hermeneutics with a philosophy that concentrates solely upon the communicative interest. He claims that an adequate philosophy needs to incorporate both approaches. It needs to deal with both causality and meaning.

In Habermas's later terminology of his *Theory of Communicative Action* (1987), work consists of instrumental action, which involves means–end activity. According to Habermas, instrumental action characterizes natural science and technology. Communicative action involves language, symbolic activity, and interpretation. Communicative action is present in the lifeworld of immediate experience (everyday conversation) and in debate and discussion within democratic politics. According to Habermas in his later writings technocracy attempts to "colonize the lifeworld" with the authority of technocracy and a technocratic way of thinking about human problems. That is, areas previously consisting of informal face-to-face interaction and traditional knowledge are replaced by the intervention of supposed experts from the social sciences (Habermas, 1987). Education and the family life are impacted by social scientists and social workers, and shift from the private realm to the domain of state intervention and social engineering by technocrats. Habermas claims that Marx and Marxism erred in attempting to reduce communicative action to instrumental action, making work the key to the reorganization of society and downplaying the communicative dimension of politics.

Habermas clearly wishes to include both the instrumental-technological and the symbolic-communicative dimensions as part of the characterization of humans. Thus his position could be seen as a compromise between the tool-making and linguistic theories of human nature, although he emphasizes the inadequacy of the tool-making theory (instrumental rationality) as a total approach for understanding human society, an approach he attributes to Marx. Unlike Heidegger, Mumford, and Arendt, Habermas does not *reject* tool-making and technology as characteristic of human being, but presents it as one of two (or three in his earlier work, including the "emancipatory interest") "equiprimordial" species characteristics of humans.

Conclusion

There has been a debate about the place of technology in the theory of human nature. The early Greek philosopher Anaxagoras claimed that manipulation of surroundings with the hands led to the growth of mind. However, Plato, Aristotle, and most of the tradition for the next 2200 years claimed that the rational, contemplative, and spiritual mind was what is characteristic of human beings and human nature. Aristotle claimed that humans are, in essence, rational animals.

In the late eighteenth and nineteenth centuries, Benjamin Franklin and others revived the emphasis on action rather than contemplation, and claimed that humans are tool-using animals. Marx emphasized work and technology in understanding human history and our present condition. Engels went further in characterizing humans as laboring animals and tool-makers.

A number of twentieth-century philosophers have reacted against the notion that humans are essentially tool-makers and that technology is the key to human evolution. Twentieth-century philosophy shifted away from mind to language and symbolism as the key to understanding rationality. Thinkers such as Mumford, Heidegger, and Arendt have emphasized language and symbol-making rather than the traditional, spiritual mind or mental substance of Plato or early modern philosophy. Part of the motivation of those who have emphasized language and meaning rather than technology as the key to understanding the special place of humans has been a pessimistic attitude to technology in general, or, in Mumford, at least the technology that has hitherto existed.

Technological optimists such as Franklin and Engels saw tools as the key to the special qualities of humans. Technological pessimists have claimed that language, not technology, is what makes humans special, and even that technology has suppressed and degraded the special capacity of humans for communication and mutual understanding of meanings. Those thinkers, such as Marx himself (as opposed to Engels and technological determinist Marxists) and Habermas, who wish to emphasize both the liberating and the oppressive capacities of technology have characterized human beings in their highest capacities, not only as active, laboring creatures, but also as creatures of creative political action or of communication.

Study questions

1   Do you see human tool-making and language as part of a quantitative continuum with animal tool-making and animal communication, or do you see a qualitative break between the human and animal capacities here?

2   What is your own judgment as to which, if either, is more fundamental to human nature, tool-making or language?

3   If we distinguish alienated labor from non-alienated labor, is Arendt's claim that Marx has more than one concept of labor or an ambiguous concept of labor still valid?

4   Does the recursiveness of human tool-making (producing tools to make tools to make tools . . .) show a genuine difference from animal tool-making, or do the discoveries of animal tool-making destroy the characterization of humans as tool-makers?

5   Does Arendt's analysis of work, labor, and action still hold in the computerized factory after the transition from the industrial to the post-industrial or information age? Does the Internet reinvent the political space that Arendt claims disappeared in the modern world?

6   Does the rise of information technology and electronic communication undermine Habermas's dualism between instrumental, technological action and communicative action? If so, how? Consider a society in which production of information rather than physical objects is the central activity. Does this blur the distinction between action and work, or does information become merely another artifact?

9

Women, Feminism, and Technology

A Woman's Work Is Never Done
I saw a man the other day,
As savage as a Turk,
And he was grumbling at his wife
And said she did no work . . .
He said: You lazy huzzy!
Indeed you must confess;
For I'm a-tired of keeping you
In all your idleness.
The woman she made answer:
I work as hard as you,
And I will just run through the list
What a woman has to do.
So men, if you would happy be,
Don't grumble at your wife so;
For no man can imagine
What a woman has to do.
Lesley Nelson-Burns (*c*.1850)

A woman's place is in the wrong.

James Thurber (*c*.1950)

Feminist philosophy of technology is part of the larger movement and project of feminist philosophy in general. Feminist philosophy started in applied ethics (Alcoff and Potter, 1993), where issues of gender with respect to abortion, child rearing, sexist language, and general issues of male power and dominance are most obvious. However, as feminist philosophy developed, feminist philosophers moved to deal with foundational issues in theory of knowledge and metaphysics. During the 1970s, as a part of so-called second wave feminism (the First Wave being the fight for women's suffrage), feminist philosophy of science and technology arose, with writers such as Evelyn Fox Keller, Donna Haraway, and Sandra Harding. Feminist approaches to the theory of knowledge are often made in contrast with in and opposition to the standard positivistic, objectivistic and technocratic ones (see chapters 1 and 3).

A number of philosophical tendencies in the latter part of the twentieth century were exploited, developed, and extended by feminist epistemologists (theorists of knowledge) and philosophers of science and technology. Criticism of logical positivism and the psychologically and socially oriented post-positivist philosophies, such as those of Thomas Kuhn (as well as Stephen Toulmin, Paul Feyerabend, and Michael Polanyi), opened issues and topics concerning social and psychological biases in science for feminist philosophers (Tuana, 1996). Likewise, phenomenological and hermeneutic approaches from continental philosophy that eventually were assimilated in US philosophy gave an entrée to feminists to introduce the role of context, personal feelings, and social situation into the philosophy of science. Criticism of uncritical celebration of technological progress and futurological fantasies of total control of nature, raised by the ecology movement as well as earlier German and English Romanticism in philosophy (see chapter 11), opened the way for feminists to point out masculine aspects of the dominating attitude to nature. Pragmatic and existential critiques of the notion of the detached observer, with the "view from nowhere," to use Thomas Nagel's phrase, were assimilated by some feminists to criticize the notion of scientific and technological objectivity (Heldke, 1988). W. V. O. Quine's pragmatic criticism of the notion of decisive refutation of theories and of a sharp distinction between the empirical and the definitional truths in science led some feminist theorists of knowledge to reject the whole foundational approach to the theory of knowledge, which bases knowledge on the intuitive apprehension of indubitable truths by individual knowers (Nelson, 1990).

There are several areas of investigation of technology in relation to women. Among these are: (a) women's generally overlooked contributions to technology and invention; (b) the effect of technology on women, including household technology and reproductive technology; (c) gendered descriptions and gender metaphors of technology and nature and their role in society.

Women's Contributions to Technology and Invention

One area of research is that showing the often underrated contributions of women to technology and invention. From prehistoric food gathering and storage to the development of the COBOL business computer language women have contributed substantially to technology (Stanley, 1995). However, what is classified as technology has often biased the account to exclude or downplay women's contributions. Even the most eloquent and influential American historian of technological systems includes few women in his recent survey (Hughes, 2004).

For instance, the anthropology of the 1960s found a unifying theory of the development of modern *Homo sapiens* in the "Man the Hunter" theory. Big game hunting was seen as central to the development of human intelligence and social cooperation. Because men were predominant in big game hunting, this meant that men were responsible for the social advance of humanity. As Ruth Hubbard asked rhetorically in another context, "Have only men evolved?" (Hubbard, 1983). During the late 1970s under the influence of feminism a number of female anthropologists developed the "Woman the Gatherer" theory or account to emphasize the contributions of women to the food supply. Some of these accounts noted that the gathering of plants, nuts, and seeds and the trapping of small game was more important to general nutrition than the occasional big game hunt.

Lewis Mumford had earlier noted how the identification of technology with machines and weapons had overemphasized the male role in invention, and the importance of container and storage technology was often overlooked (we noted this in our discussion of animal technology in chapter 8). Mumford noted that although the extension of the leg in transportation devices and the extension of the arm in projectiles have been emphasized, a kind of prudery has led historians of technology to ignore the extension of the womb and the breast as storage or incubation devices (Mumford, 1966, pp. 140–2; Rothschild, 1983, p. xx). During the Middle Ages the invention of the quern or hand-cranked grain mill, a part of women's work, introduced the crank to mechanics (White, 1978).

In more recent centuries the assumption that women have not been inventors, as claimed by Voltaire (Stanley in Rothschild, 1983, p. 5), has led the stories of women inventors to be ignored, covered up or misinterpreted. Often it is assumed that if women made any inventions they concerned "women's work," i.e. housework. One women inventor of a design for a river dam had her patent application misinterpreted as a design for a "dam" in a kitchen sink! As in other areas, such as literature, the production or invention of women was attributed to their husbands. The frequently disputed case of Catherine Green's contribution to the invention of Eli Whitney's cotton gin is a famous example. Green probably suggested the use of a brush to remove the cotton lint that stuck to the teeth of the cylinder (Stanley, 1995, p. 546). Emily Davenport made crucial contributions to Thomas Davenport's small electric motor. Ann Harned Manning jointly invented a mechanical reaper with her husband William Manning before McCormick invented his, but it is the husband William who generally is given credit.

Technology and Its Effects, Particularly on Women

Two areas that most obviously have had effects specifically on women are household technology and reproductive technology.

Household technology

During the twentieth century a number of mechanical inventions changed the nature of housework. The washing machine, the vacuum cleaner, the gas, electric, and microwave ovens, and frozen food are examples. Indoor plumbing and the automobile also had great effects on household work and the allocation of time.

Surprisingly, the introduction of these household devices has not shortened the hours spent by house workers and mothers (Cowan, 1983). For upper-class women the decline of use of servants offset the greater efficiency of the washing machine, vacuum cleaner, and oven. For less affluent housewives the increased efficiency of these household devices increased output but did not decrease work. The washing machine saved time and effort over hand washing, but the use of both hired laundresses and professional laundries declined. The greater efficiency of the washing machine also led people to change their clothes and hence wash their clothes more frequently. The vacuum cleaner led to houses being much cleaner, but house size grew during the suburban boom of the 1950s. There were more areas to clean. Thus clothes and houses were both much cleaner, but cleaning houses and washing clothes was more frequent and extensive. The new ovens and prepared and frozen foods decreased food preparation time, but other activities took its place. Furthermore, the disappearance of many of the physically exhausting and obviously skilled activities involved in food preparation and cleaning often led to a decline in the respect husbands held for their spouse's housework. The widespread myth from at least the 1950s was that house-wives had hardly any work to do.

The availability of the automobile changed the activities of housewives. Previously milk, bread, ice, and most groceries were delivered to the house-hold door and physicians made house calls. With the automobile trips to the store and to the doctor became more frequent. Indirect effects of the develop-ment of the automobile system led to the spread of the suburbs, the growth of malls and supermarkets, and the decline of local mom and pop stores, all increasing the amount of travel needed for food shopping. The decline of public transportation, in part due to the dominance of the auto-mobile, also means larger demands on the auto for transportation. Much time is spent transporting children to and from various activities.

Despite the improvements in household technology over three decades in (what used to be called) the "actually existing socialist" or Soviet style coun-tries of Eastern Europe, the answer to the question "Does socialism liberate women?" seemed to be a qualified "No." In the Soviet Union, although women early worked full time, they were also expected by their spouses to do all the housework. There were some early experiments with collective laundry, and even cooking, but these were not sufficiently widespread or long lasting to ease the burden on women (Scott, 1974).

In the development of household technology there is a gender split between the designers (almost entirely male) and the consumers (mostly female). The central house vacuum is sold in Sweden by appealing prim-arily to the female user but also to the male as purchaser, lint dumper, and repairperson (Smeds et al., 1994). The microwave oven in Britain began as a "gee whiz" gizmo sold in stereo and electronics shops. Only later did it shift to be treated as an ordinary household appliance sold in appliance shops. In this latter placement sales techniques are focused on women's fears of com-plex technology and of the danger to health from insufficiently cooked items causing food poisoning (Ormrod, 1994).

Reproductive technology

A second area of technology that has obviously influenced women's lives is reproductive technology.

During the first years of second wave feminism in the early 1970s Shulamith Firestone's *The Dialectic of Sex* (1970) proposed that only separating women from pregnancy and childbirth through artificial wombs could achieve full equality of women. This technological fix approach was soon rejected by most feminists, who tended to emphasize women becoming more involved in and in control of their pregnancies. Later feminists, who emphasized the less desirable aspects of artificial reproduction technology as a means of power of male physicians over women, also rejected it.

Contraceptive and abortion technology is about delaying or avoiding pregnancy. Artificial insemination, embryo transplant, and other new reproductive technologies are about achieving pregnancy. Feminists have been concerned to extend the availability to women of contraception and abortion so that women are in control of whether and when they become pregnant. Feminist critics have focused on the alleged lack of concern for women's health in contraception research and development and the relative lack of research into chemical forms of male contraception. The case of Depo-Provera is an example. Depo is a contraceptive injection that lasts for three months. Because it doesn't involve the need to remember to use a physical contraceptive or to frequently take a pill, it was the contraceptive commonly given to Third World women, to aborigines in Australia, Maori in New Zealand, and women of color in Britain. It is claimed that US AID (Agency for International Development) channeled funds to the International Planned Parenthood Federation to distribute Depo worldwide. Data from studies in New Zealand by its manufacturer, Upjohn, were sent to company headquarters for statistical analysis and not released publicly. Critics have claimed that the published claims concerning the drug downplay side-effects such as cancer and bleeding (Bunkle, 1984).

An apparently surreptitious campaign during the 1950s and 1960s of massive sterilization of poor, Hispanic women in Puerto Rico without informed consent of the subjects is another example of direct reproductive control of Third World women.

The proponents of reproductive technology have emphasized the increase in freedom of choice that the new reproductive technologies have offered women. Contraception, *in vitro* fertilization, embryo implantation, and genetic screening are among these technologies. The ability to prevent pregnancy, the ability of previously infertile women to bear children, and the ability to screen for and abort fetuses with genetic defects are presented in terms of extended capability and free choice. Feminist critics of the new reproductive technologies, on the other hand, have noted that the new possibilities have imposed subtle pressures and constraints on women. Infertile women are expected to make use of the new reproductive technologies to be able to reproduce. Women are expected to screen for and abort "defective" fetuses. A woman who does not make use of genetic screening or who elects to give birth to a child with a genetic defect is considered morally derelict by those who accept abortion and the new technology (Rothman, 1986). Critics in the disability rights movement have also noted that the eagerness to eliminate "defective" embryos shows society's negative attitude to the disabled. Racial, ethnic, and class issues also enter into eagerness to abort the potentially disabled (Saxton, 1984, 1998).

Radical feminist critics of the new reproductive technologies have claimed that they are a means for the mostly male physicians to control the one human act (pregnancy and childbirth) that men are unable to do. In early societies there was a religious mystique about the reproductive powers of women. In the Renaissance and early modern period one of the dreams of the alchemists was to produce the homunculus or little person by purely chemical means. This would allow male alchemists to achieve the one ordinary human task of which they had previously been incapable. Some feminist theorists of technology have seen modern reproductive technology as a fulfillment of this age-old dream of male capability and power. Feminist critics have seen contemporary genetic engineering and test tube babies as a fulfillment of the homunculus fantasies of alchemists, such as Michael Maier's *Atalanta Fugiens* (1617, emblems 2–5, 20, see Allen and Hubbs, 1980).

Maier, with his highly sexually charged and often misogynist symbolism, was a favorite alchemist of Isaac Newton, who rejected contact with women (Dobbs, 1991, n123). Maier was evidently also involved in the conquest of Native Americans. On his visit to England he was an associate of at least three members of the Virginia Company, planning to settle America, two of whom associated this project with alchemical ideas, including those of Maier. Maier's own *Atalanta* may have been, in turn, partly inspired by anticipation of a colony in Virginia (Heisler, 1989).

The replacement of female midwives with male surgeons in the early modern period was a shift in who had knowledge of and power over childbirth. The early surgeons' takeover was facilitated by the development of a simple technology, the medical forceps, introduced in the 1730s. Despite the fact that in this early period the surgeons killed more than they cured, often overusing the forceps, damaging both infant and mother, they were able to present themselves as experts more worthy of respect than the midwives (Wajcman, 1991).

Later, more technologically sophisticated and successful developments in obstetrics, including anesthetics, along with the move of birthing from the home into the hospital, completed the medicalization of pregnancy and childbirth. Pregnancy became pathology rather than a natural process and part of human life. Caesarian sections were performed more often than necessary (sometimes for the convenience of the physician, who then did not have to wait during hours or days of labor). This, along with induction of childbirth and episiotomy, led to the management and control of pregnancy, and control shifting from the mother and midwives to the male physicians.

Within the new reproductive technologies, techniques of sex selection have the most direct and obvious effect on gender discrimination. Because of the valuation of male offspring over female offspring in traditional societies, China and India have been involved in extensive selection of male embryos and abortion of female embryos.

One area that shows the complexity and ambivalence of the new technologies is ultrasound imaging. Although this has become a routine part of medical management of pregnancy, several studies have cast doubt on the value of routine ultrasound imaging on the improved health of fetuses and offspring. The ultrasound image allows the physician to be in control of knowledge of pregnancy that is superior to that of the mother. It also shifts the kind of knowledge involved. Traditionally the mother's feeling of quickening and of kicking in the womb was the means of sensing the fetus. This has been replaced by the visual imagery of the ultrasound image. Many writers since Friedrich Nietzsche (1844–1900) have noted that visual perception is a more detached, "distanced" kind of perception than that of the other senses (Jay, 1993). During the past few centuries visual perception has been given priority over other senses, such as touch and smell. The visual is linked to the geometrical and objective account of the world of modern science. The mother's feeling of the fetus is part of the mothers' own bodily sensations, while the visual image is an image from the "outside." The attention is on the screen, not on the mother's body. The ultrasound images show the fetus as if isolated in space, ignoring the bodily medium in which it is suspended, giving the impression of independence of the fetus from the mother. They are similar to the floating or flying fetus in the final scene of *2001: A Space Odyssey*, thereby eliminating the presence of the mother, and associating the fetus with high medical technology rather than human gestation (Arditti et al., 1984, p. 114).

Ultrasound has been claimed to change the very notions of "inside" and "outside" with respect to mother and fetus. One writer calls the ultrasound a kind of "panoptics of the womb," along the lines of Michel Foucault's

(1977) notion of universal surveillance. *Life* magazine, in one of the earliest popular accounts of ultrasound, claimed it works "precisely the same way a Navy surface ship homes in on an enemy submarine" (1965, quoted in Petchesky, 1987, p. 69). One physician writing in a major medical journal used phrases very much in the Baconian mode (see the section on metaphor below), such as "the prying eye of the ultrasonogram," "stripping the veil of mystery from the dark inner sanctum," and "letting the light of scientific observation fall on the shy and secretive fetus." Indeed, Harrison calls the fetus a "born-again patient" (quoted in Hubbard, 1983, pp. 348–9). The anti-abortion movie *Silent Scream* makes use of medical imaging (with acceleration of the speed of the images, and an unrealistically large model of the fetus) to persuade the viewer that the fetus is a person. In several US states, laws have been proposed to require viewing of ultrasound images of the fetus by women seeking an abortion. The ultrasound scan, a technology with numerous undoubted medical advantages, can be used as a weapon to present the "fetus *in situ*" (now treating the woman as a mere lab vessel), as if independent of the mother bearing it. The imaging can be, alternatively, an extension of the age-old voyeurism of the male "gaze," with women objectified and depersonalized, similar to the images of pornography in this respect.

Workplace technologies and women

Not all the technologies that have affected women are specifically oriented toward women's traditional roles as mother or as housewife. Industrial technologies have affected women's occupations. One debated example is the typewriter in relation to women entering the clerical workforce.

Earlier accounts of the development of the role of secretary as populated by women rather than men by the 1920s tended toward a technological determinist account, linking the shift in gender identification of the job to the rise of the typewriter. It has been noted, however, that Japan developed a largely female clerical force without the use of the typewriter. Indeed, the increase of women in secretarial and clerical jobs in the USA began during the Civil War, a decade and a half before the typewriter was more than a rarity. The design of the typewriter as a cross between a piano and a sewing machine seemed appropriate to women. Women had occupied the low-wage job of hand copying, and the typewriter was widely used for copying. The development of stenographic writing to take dictation led to the training of women typists in stenography, which was originally a male field. As women were trained and credentialed in typing and stenography at secretarial schools, stenography became women's work. For small businessmen the power over a woman assistant gave a sense of authority and prestige. Given the low wages of female copyists, and their successors, the typist-stenographers, the job became less attractive to men. The typist-stenographer role was a bridgehead to being hired in other clerical jobs, leading to the new characterization of clerical work as female rather than male by the 1920s and 1930s (Srole, 1987).

Technology as "Male," Nature as "Female": Metaphors of Nature and Technology

There is a rhetoric of "Man and Technology" and of "Man's Domination of Nature." It has been noted by a number of feminist writers that nature is generally portrayed as female, as in Mother Nature or virgin lands. Scientists and technologists are generally portrayed as male. This convention goes far back in time. Ecologists like to cite the earth goddess Gaia, and James Lovelock has used her name to designate his theory of a self-adjusting biosphere, including both chemicals, such as atmospheric gases, and organisms.

With the rise of early modern science in the sixteenth and seventeenth centuries there was, according to Carolyn Merchant and others, a downgrading of the status of the earth mother (Merchant, 1980). This was associated in part with the rise of exploitation of nature. Miners who lacked reverence for the earth, and saw it not as a living thing but as an inorganic, lifeless mass, would be less restrained in their excavations and extraction.

Much early thought on nature, such as that of the early Greek natural philosophers, treated matter as alive (hylozoism). However, in the seventeenth century, the mechanistic view in its original form treated matter as wholly passive and inert. Aristotle had treated matter as passive in analogy to the female and form as active in analogy to the male. However, the early atomists and mechanists further emphasized the passivity and deadness of matter. Thinkers in the hermetic tradition treated the forces and powers of matter as active, but in opposition to them the early mechanists such as Descartes denied any active powers to matter. Newton realized that, contrary to Descartes, forces were necessary to produce an effective theory of physics, but Newton denied that the forces resided in matter. He claimed that they were a separate form of "active principle."

Although Newton borrowed ideas from alchemy in developing his concept of force, he degraded the female principle in theory of matter and had a pathological aversion to social contact with women. Newton became angered at his friend John Locke and refused to speak to him because Newton thought that Locke was attempting to involve him with women. One of his alchemical metaphors was "ye menstruum of [your] sordid whore." More than other alchemists, Newton was fascinated by "the net," a chemical associated with the net by which Vulcan traps Mars and Venus when they are caught *in flagrante delicto* (Westfall, 1980, pp. 529–30, 296; Dusek, 1999, p. 185).

Not only were feminine qualities downgraded or dismissed in the theory and metaphors of matter, but the new natural philosophy was also seen as masculine. Henry Oldenburg, the correspondent for the Royal Society, argued for a "masculine philosophy." Joseph Glanville, a historian of the Royal Society, also demanded a "manly sense" and advocated avoiding the deceit of "the woman in use" (Easlea, 1980, p. 214). Oldenberg not only advocated a "manly philosophy" but demanded that "what is feminine . . . be excluded" from the philosophy of the Royal Society. Thomas Spratt, who wrote an early history of the Royal Society, likewise identified the intelligence of the crafts as masculine (Keller, 1985, pp. 54, 56; Dusek, 1999, pp. 128–36).

Francis Bacon (whom we have encountered both in the discussion of the philosophy of science as proponent of inductive methods in science and in the discussion of technocracy as forerunner of technocracy and booster of the value of natural knowledge in the prosperity of society) used a variety of gender images for the relationship of the (male) scientist with (female) nature. He used the image of marriage. He also used the imagery of voyeurism and seduction of nature. He associates the male investigator of nature with the probing of secret places and the forcing of nature to reveal her secrets. The quotation in Bacon that has caused the most controversy is:

> For you have but to follow and as it were hound nature in her wanderings and you will be able when you like to lead and drive her afterwards to the same place again. Neither ought a man make a scruple of entering and penetrating those holes and corners when the inquisition of truth is his whole object. (Harding, 1991, p. 43)

Feminist writers have associated this passage with the fact that Bacon dedicated his work to King James I, who was active in investigating and persecuting witches. This passage is notorious, but there are many other gendered treatments of nature as slave and object of capture (Farrington, 1964, pp. 62, 93, 96, 99, 129, 130).

Feminist critics of Bacon have seen the relationship of male experimenter to nature as a kind of forceful seduction, verging on date rape. Allen Soble

(1995), in his defense of Bacon, identifies the relation of investigator and nature with marriage and argues that marital rape was legal in Bacon's day (and for a long time afterward). He also claims that one cannot find a "smoking gun," by which Bacon directly linked experiment with the rape of nature, or associated the investigation of nature with the torture of witches. Nevertheless, the passage cited above and many others show that Bacon places his consideration of the investigation and manipulation of nature within a gendered context.

One may say against this analysis of the metaphors of early modern science and technology that they are not essential to the science and technology themselves. The experiments, laws of nature, and mechanical inventions stand by themselves and the metaphor is only exterior decoration. However, the imagery of male gender in the investigation of nature is so pervasive, continuing to our time, that one may argue that such images and metaphors have an effect on the image of science and technology that plays a role in the recruitment and motivation of scientists and engineers.

Evelyn Fox Keller (1985) and others, using psychological object relations theory (Chodorow, 1978), have claimed that the very norm of detachment and objectivity in science and technology is associated with the male model. According to object relations theory, boys have to break with their mothers in the formation of their identity in a way that girls do not. The masculine stereotypes of lack of emotion, detachment, and objectivity fit with the image of science. These popular images of science and technology influence the recruitment of students into the fields. Girls in middle and high school who have talents for science and technology are discouraged from pursuing advanced technical subjects by the popular images of scientist and engineer. These images of nerd, on the one hand, or aggressive controller of nature, on the other, conflict with the "feminine" personality traits that society encourages girls to develop. Girls also fear that even excessive intelligence or talent for technical subjects will discourage boys' interest in them.

David Noble in his A World without Women (1992) has emphasized that priests and monks who were not married and supposed not to have sexual relations with women carried out medieval scholarship. The academic world grew from the medieval universities, such as Oxford, Cambridge, Paris, Padua, and elsewhere, which had clerical origins and solely male inhabitants. Women were not admitted to major universities until the nineteenth century. Many eminent US men's undergraduate universities, such as Yale, did not become coeducational until around 1970. Women could not pursue advanced work at the major universities until the nineteenth century in

Russia and the twentieth century in parts of Western Europe. When the mathematics faculty of Göttingen, Germany, resisted accepting one of the world's leading experts on algebra because she was a woman, the mathematician David Hilbert asked in exasperation, "Is this a university or a Turkish bath?" (Reid, 1970). The exclusion of the feminine was not simply in imagery and psychology but in the institutional structure of academic science and technology.

The debate concerning the issue of relevance of metaphors to science and technology is related to the positivistic and post-positivistic philosophies of science, as well as to the definition of technology in terms solely of hardware or of rules, versus the technological systems definition, which includes social relations within technology (discussed in chapter 2).

According to the positivistic view, science consists of the formal deductive apparatus of the theory and the observational data. Models, metaphors, and heuristic guidelines for discovery are not part of the "logic" of science but part of its "psychology." For the most part they are in the context of discovery, not of justification. Only the logic of explanation and confirmation in the latter is significant for knowledge. However, some philosophers of science, such as Mary Hesse (1966), Rom Harré (1970), and Marx Wartofsky (1979), have argued that models are an important part of scientific theories and explanations.

Social historians of science and technology and sociologists of scientific knowledge have claimed that the broader social and cultural images and metaphors play a role in the acceptance and spread of scientific theories. For instance, Darwinian natural selection was stimulated by Malthus's theory of the economics of human overpopulation (which also was a trigger for the independent co-discoverer of the theory, Alfred Wallace) and by Quetelet's work in social statistics. The acceptance of Darwinism was aided by the resemblance of the theory to that of the competitive capitalist free-market economy (Gould, 1980; Young, 1985).

A more controversial example is that the model of the universe as one of atoms moving and colliding in empty space, with no natural up or down, mirrored the competitive, capitalist, free-market economy, replacing the hierarchical Aristotelian worldview of the Middle Ages in which things had their natural place, and the levels of the hierarchy were levels of value (Brecht, 1938; Macpherson, 1962; Rifkin, 1983; Freudenthal, 1986).

Similarly, the technical or hardware understanding of technology would exclude the imagery and cultural values that inventors or users of technology might associate with the technology as not really part of the technology.

The technological systems definition of technology includes the social organization of the maintenance and consumption of the technology. Hence images and metaphors that motivate inventors or make the technology attractive to consumers have a role in the technology. If, for instance, the imagery of making a "second creation" of life and the taking over by male scientists and medical men of the power and mystery of human reproduction motivates molecular biologists, genetic engineers, and reproductive physicians, then it is a part of the social system of that technology and culture of that technology.

Variant Feminist Approaches to Theory of Knowledge in Changing Science and Technology

Sandra Harding (1986) classified approaches to science in a way that has relevance to technology as well. The position closest to the traditional and widespread theory of scientific knowledge is **feminist empiricism**. Feminist empiricism aims to reform science and its technological applications – for instance, in medicine – by correcting bad science. It accepts standard empiricist or even positivist accounts of the nature of scientific knowledge, claiming that what is wrong is simply bad science and false claims about women.

Much of feminist science criticism has been directed at biological theories of women's intellectual inferiority and lack of motivation as explanations of women's lack of participation in science and technology. Numerous accounts of women's lack of mathematical ability are allegedly based on psychology and brain science. The accounts have shifted over time but have managed to maintain claims of a lack of female abstract reasoning ability. With the discovery and popularization of differences in the two hemispheres of the brain, right associated with intuitive and holistic grasp and left associated with language and formal thinking, women (but also non-Western people in general) were initially claimed to be right-brained, intuitive, and non-logical. This image fit with popular stereotypes. After it became obvious that girls' language development led boys' the story changed. The recent version is that women are left-brained, skilled in language, but this is now used against women's abstract abilities. It is claimed that mathematical, spatial, and geometrical skills are associated with the right brain, and that boys are right-brained with superior spatial skills. This neglects purely linguistic and non-spatial areas of math, as in logic and computer science. The extreme extension of this approach is the speculative claim that men have a "math gene" that women lack. This is based on differences in SAT scores. There is no genetic evidence for the supposed sex-linked "math gene" (Moir and Jessel, 1992; Hammer and Dusek, 1995, 1996).

The alleged discovery that women have larger connective tissue (splenium) between the two parts of the brain has been used to claim that women are less able than men to separate thought and emotion. Biologist Anne Fausto-Sterling (2000) among others has pointed out the small samples and non-replicable nature of the studies that make these claims. Feminist studies have also criticized the scientific quality of studies in sociobiology and more recently evolutionary psychology that give a supposed evolutionary basis for the claims about women's lack of abilities in abstract or technical fields.

Feminist empiricists believe that an honest and accurate use of traditional scientific methods will undermine biases against women in science and technology. As feminist empiricism exposes more and more bias in the descriptions of human and animal behavior, one is led to question the extent to which scientific method, traditionally applied, is sufficient to expunge sexist bias. If the leading peer-reviewed journal *Science* can publish an article on "transvestitism" among hanging flies (Thornhill, 1979), even though the insects do not wear clothes, one wonders whether traditional peer review can function to correct for bias.

Other feminist approaches claim that more substantial changes in our usual accounts of science and technology are needed. Feminist **standpoint theory** is a more radical approach. The structure of the theory is based on an aspect of Marxist theory. Georg Lukács (1923) claimed in his early work that the standpoint of the worker, central to the process of industrial production, but also oppressed and alienated, gave a privileged access to knowledge denied to the comfortable and detached capitalist owner. The worker as "self-conscious commodity" had direct, personal insight into reification of the self that the capitalist or professional did not. Feminist standpoint theory makes a similar claim for the position of women as central to society's reproduction but oppressed. Unlike men, who generally take for granted and do not notice the gender exclusions and gender discrimination built into the structure of the technical community, women are forced to become aware of the biases directed against them.

Post-Kuhnian philosophy of science emphasizes the extent to which guiding assumptions over and above the formal theories and bare observational data function. Paradigms include ideals of theory, as well as an image of nature. Feminists have claimed that stereotypical images of science and technology, as control and manipulation of nature (rather than, say, understanding and cooperation with nature) and as analytical breakdown and reduction of systems to their simplest parts (as opposed to recognition of holistic effects and emergent levels of systems), have gender bias behind them. Feminist standpoint theory claims that women's situation forces them or at least makes them more likely to become aware of these biases than are men.

One objection to feminist standpoint theory, as it is to the Marxist version of standpoint theory originated by Lukács, is to question whether oppression and pain themselves are automatic tickets to objectivity. They might produce their own distorting biases.

Some feminists argue, as do many **ecofeminists** (see chapter 11) and feminist participants in disarmament or anti-nuclear movements, that women's nature, including their involvement in childbearing and childrearing, makes them more apt than men to be concerned with the existence of future generations and the preservation of the planet. On a broader scale, feminists committed to a notion of women's nature, as well as ecofeminists in general, claim that women are "closer to nature." Whether attributing it to male and female natures or to the power structure of society, ecofeminists claim that there is a connection between patriarchy or male domination of society and values of domination and control over nature.

Male nature is claimed to tend to abstract and reduce, to "murder to dissect," while women respect the integrity, complexity, and fragility of natural systems. Another claim is that women are cooperative and non-hierarchical, while men are predisposed to be competitive and hierarchical. It is claimed that many technological networks and systems reflect the centralized control and hierarchy of a male-dominated society and that greater participation of women would lead to a more decentralized and democratic technology.

One irony of the position that holds that there is a women's nature and a men's nature is that it parallels the claims of the sociobiologists and other biological determinists who use similar accounts to claim that women are unsuited for technology because of the same characteristics that the women's nature theorists attribute to them. The difference is that the women's nature theorists positively value the irrationality and emotionality that traditionally has been regarded as inferior to male rationality. The women's nature theory also repeats the imagery and metaphors of much of the rhetoric of the scientific revolution, with science as "male" and nature as "female," and the relation of scientist to nature as that of man to woman. While the sociobiological "anatomy is destiny" theorists claim that women's lack of ability to totally detach their abstract thought from their emotions, and their supposed lack of aggressive competitiveness, make them lack scientific and technological ability, the women's nature theorists claim that these very attributes will either eliminate technology as we know it or lead to a more humane and beneficial science and technology.

Opposed to the theory of women's nature are **postmodernist feminism** and the anti-essentialist claim that gender is socially constructed. Postmodernism is a diverse movement of the last decades of the twentieth century that among other things denies that there is the possibility of a complete system of knowledge or a metaphysical account of ultimate reality (see the sections on post-industrial society, media, and postmodernism in chapter 6). Postmodernism would deny that feminist standpoint theory could lay claim to the possession of the *true* standpoint.

Postmodernism is a relativism that claims that there are a variety of standpoints with equal claim or lack of claim to the truth. Postmodernism also denies that there are essences (see chapter 2). Words and definitions are arbitrary. There are no natural classes of things or natures of things. In particular, postmodernist feminism denies that there is a women's nature. Gender is socially constructed (see chapter 12). That is, the personality characteristics that a society attributes to women and to men do not reflect a real nature of women or of men but are a product of the society itself.

Another feature of postmodernism is the denial of a unified identity of the self. Postmodernist feminism emphasizes the extent to which individuals are identified with a number of groups. Women are not simply women but women of a certain race and class. Because of this the female "essence" cannot be used to characterize a woman's political and social place.

Donna Haraway is a postmodernist feminist who has greatly contributed to the construal of science and technology. In her *Primate Visions* (1989) she shows how the interplay of gender and race affects the portrayal of the great apes in science and in popular discourse. In her "Manifesto for Cyborgs" (Haraway, 1985, 1991) she develops the category of the cyborg, a combination of human and machine, to undermine the dualities of the human and the mechanical and to reject the notion of a human essence. The cyborg originated in technological speculation concerning long-range space travel and in science fiction, but Haraway and others since have claimed that in fact this interpenetration or inseparability of the human and the technological is characteristic of our condition. In contrast to the romantic and the essentialist feminist opposition of the natural to the technological, the cyborg shows the two as inextricably intertwined. This cyborg breaks down the line between human, animal, and machine. Genetically engineered organisms and even, in Haraway's more recent work, companion animals such as dogs blur the boundary between the natural and the artificial. Haraway's use of cyborgs to erase the traditional boundaries that humanism and essentialism have erected is similar in many respects to Bruno Latour's use of technology–human hybrids or "quasi-objects" to undermine the opposition between traditional positivist objectivism and social constructionism (Latour, 1992, 1993).

An example of such a nature/culture hybrid and its resonance in both science and quasi-religion is the dolphin. *Cosmodolphins* (Bryld and Lykke, 2000) is a feminist cultural studies work, utilizing Heidegger, Cassirer, various postmodern thinkers, and feminist philosophers, that reveals the ambiguities and ironies of contemporary attitudes to nature and the universe. High technology projects for space travel and extraterrestrial communication mirror quasi-religious beliefs about higher intelligences in outer space and New Age visions of harmony with nature. Dolphins have long been considered intelligent and worthy companions of humans. They have also been thought to be models for communication with alien intelligences by leaders of the SETI (Search for Extraterrestrial Intelligence) community. The experiments of neuroscientist John C. Lilly started as traditional, cruel, constraining and invasive neurophysiologic probes, but led Lilly to believe he was actually speaking with dolphins. Lilly himself moved to sensory deprivation tank immersion and LSD to attempt to achieve dolphin-like consciousness. His experiments received favorable popular attention worldwide, and many non-scientists today believe that human–dolphin conversation has been achieved.

The popularity of dolphin and orca performances at aquaria and aquatic parks show the popular fascination with cetaceans. Dolphin imagery has spread in advertisements for telecommunications and computer software. Carl Sagan, the solar system scientist of extraterrestrial life and TV popularizer of science, met a waitress at a Virgin Islands restaurant who then became Lilly's assistant, and soon came to head the research while Lilly was immersed, drugged, and incommunicado. Another leading physicist and popularizer of science, Philip Morrison, veteran of the Los Alamos A-bomb project and MIT professor, early advocated communication with dolphins as a bridge to communication with extraterrestrials. New Age occultists and NASA scientists are shown to share certain mythic and quasi-religious attitudes to dolphins. The claimed objective detachment and impersonality of science becomes mixed with religious awe and a desire to fuse with the cosmos. The masculinist image of cosmic domination through space travel becomes entangled in an ironic dialectic with the feminist and ecological utopia of unimpeded communication and cooperation with nature. The *Cosmodolphins* authors' project is to undermine the masculine/high technology versus feminine/occultism split. The cosmic fantasies of the proponents of the space program closely resemble those of spiritual ecofeminists and New Age occultists (Bryld and Lykke, 2000, p. 36).

The strange association of dolphins with extraterrestrials by both New Agers and mainstream scientists parallels the role of apes, both as symbol of idyllic nature and as agents of space exploration. Donna Haraway, in her chapter "Apes in Eden: Apes in Space" (Haraway, 1989, pp. 133–9), uses the *National Geographic* image of the hand of ape and woman (Jane Goodall) entwined (which also appears on the dust jacket of the book) as somewhere between holding hands and the touching of fingers of God and Adam in Michelangelo's Sistine Chapel ceiling. Surprisingly, in the contrast of apes in space with those in Eden, Haraway doesn't mention Tom Wolfe's portrayal of the jealousy of the human Apollo astronauts for the space apes in *The Right Stuff* (1979) or the movie made from it (1983).

Apes as well as dolphins are objects of human communication (see chapter 8), but are also physical objects to be manipulated and exploited by the military, as the dolphins are used to find undersea mines. Just as apes are often implicit stereotypic stand-ins for African or African American humans, from Tarzan to sociobiological studies of male inner city aggression (Goodwin, 1992; Breggin and Breggin, 1994; Wright, 1995; Sherman, 1998), so the ape–female relationship from King Kong to African primate researchers Jane Goodall and Dian Fossey can stand in for sexual as well as interracial relationships in popular culture. A recent, less literarily elegant example of apes as extraterrestials, standing in for issues of gender and race, is the kiddy cartoon Scooby-Doo video *Space Ape at the Cape* (2003), set in Cape Canaveral, prior to a space launch with a trained ape. A rapidly growing supposed alien is thought to be from an extraterrestrial egg, but turns out to be an African American female researcher covering up the failure of her SETI science project, dressed in an ape suit.

Conclusion

Feminist philosophy of technology deals with a variety of issues with a variety of approaches. It counters the traditional downplaying of women's historical and contemporary role in technology, both as users and as innovators. It investigates the aspects of technology that particularly impact women in their traditional roles as mothers and homemakers, through reproductive technology and household technology. It also examines broader issues of masculinist attitudes to technology and nature, in which male technology manipulates and dominates female nature. Feminist philosophers of science and technology have been particularly sensitive to the metaphors and cultural resonance of technology that are often dismissed by the technologists themselves. Feminist philosophy of science and technology does not speak with one voice. Investigations range from empirical criticisms of biological, psychological, and technological claims about women's nature or (lack of) ability, to alternative visions of how science and technology might be if women had more say in the directions of research and development. Postmodernist feminism investigates the inadequacies of the very dichotomy between humans and nature that lies at the basis of much traditional philosophy of technology.

Study questions

1   Is the urge to dominate and control nature particularly male? Is it an outgrowth of capitalism? Of the Judeo-Christian tradition? Is it human nature?
2   Do you think the low representation of women in the fields of physics and physical engineering will change greatly in the near future? Why?
3   Does the notion that technologies have a "valence," are structured in such a way as to be easier to use for the purposes of some groups and not others (such as men and not women), make sense, or is technology intrinsically neutral with respect to uses and users?

10

Non-Western Technology and Local Knowledge

Most discussion of technology concerns itself with science-based contemporary technology. The science at the basis of this technology is Western, stemming historically from the scientific revolution in Europe in the seventeenth century. Most of this technology is generally claimed to stem from the European industrial revolution of the eighteenth century. (Recent historians, however, have been finding more and more non-Western influences on Western science and technology, particularly from the Arab world and from East Asia, but also from the pre-conquest Americas and Africa.)

Non-Western technology, either from before the scientific revolution in ancient and medieval cultures, or more recent technology, but not based on Western science, raises a number of significant issues. One is the claim of Western science to be universal, applicable to all times and places. The mainstream Western view has been that non-Western science is, at best, an imprecise and vague formulation of narrowly applicable rules of thumb that are sub-cases of the more precise and general laws of Western science, or, at worst, superstition. Some scholars of non-Western science and members of the science and technology studies community who take an anthropological approach to science have challenged this view. They claim that Western science, itself, is a kind of "local knowledge," appropriate for the laboratory, just as non-Western science and technology is appropriate for its own environment and community.

A related, but perhaps even broader, issue concerns the usual contrast between "rational" Western thought and "irrational" or at least "non-rational" thought in indigenous pre-literate cultures. Nineteenth- and early twentieth-century anthropology contrasted "primitive thought" with modern, Western, scientific thought. More recent anthropologists and students of technology studies influenced by them have claimed that indigenous thought and technology are not "primitive" and that Westerners are only "rational" in highly limited professional contexts of science and technology. Most Westerners most of the time, anthropologists claim, including scientists and technologists outside of their specialties, think like so-called "primitive natives," relying on myths, cultural stereotypes, and loose analogies.

A third, less abstract, issue is that of the superiority or even the appropriateness of Western technology in developing nations. In the nineteenth century and through most of the twentieth the prevailing view was (and still is for many) that developing nations should imitate the technology and organization of developed nations and import Western technology to replace their own. More recently, examples of failures of implanting advanced Western technology in the environments of developing nations have suggested that less complex and difficult-to-service technology is needed. This technology is called "appropriate technology" or "intermediate technology." It is "appropriate" to less developed nations, or is "intermediate" between indigenous technology and advanced Western technology. Another claim defended by many students of contemporary non-Western and indigenous technology is that local, indigenous technologies of non-literate cultures often have great usefulness and applicability to their environment. These critics further claim that in the past Western colonial powers often dismissed indigenous technology and logical knowledge, only to replace it with techniques less efficient and effective in the tropical, arctic, or other environments. Furthermore, present Western aid projects are mistaken to dismiss the local, traditional technology and replace it with Western technology inappropriate to the environment. For example, in southern Africa, a tire factory produces tires for export to Europe. It employs local citizens, but automobiles are rare in the area. A bicycle factory would supply a mode of transport actually used by many local residents.

Local Knowledge

One of the contrasts between scientific and pre-scientific or traditional knowledge is the contrast between the local nature of traditional or indigenous knowledge and the universal nature of scientific knowledge. Indigenous, traditional knowledge is often oral in nature and conveyed by an apprenticeship in skills. The skills and lore are often secret, or at least not public. The standard view is that scientific knowledge is, as Ziman (1968) calls it, "public knowledge" (recall Merton's "communism" of data-sharing in his norms of science discussed in chapter 1). Indigenous knowledge includes detailed knowledge of the local environment, both social and biological, and has been called "local knowledge." In contrast, scientific knowledge is generally considered to be universal in at least three senses. First, scientific *laws* are logically, spatially, and temporally universal. Second, scientific knowledge can be *applied* anywhere in the universe. Third, Western science-based technology has a *geographic universality* of applicability. Any society can use it in any environment. In contrast, indigenous technology depends on locally handed down skills and on a particular, local, environmental situation.

This view of the contrast has been challenged by the science and technology studies (STS) movement of recent decades. STS is an interdisciplinary area including the sociology of scientific knowledge, the anthropology of science, literary studies of science, studies of scientific rhetoric, and approaches to the history of science influenced by these approaches. But the tone of STS is not simply a product of interdisciplinary efforts. The field, in contrast to traditional history of science and sociology of science, is often sympathetic to postmodernism in its rejection of positivistic approaches to scientific and technological objectivity and skepticism concerning the progress of science and technology (see box 6.3).

According to much of the literature of STS, science itself is a form of local knowledge. Consequently, Western science-based technology is also a special form of local knowledge. On this view, the locality of scientific knowledge is the laboratory. In the special, artificial conditions of the laboratory "effects" are produced (Hacking, 1983). The results of experiments involve the purifying and processing of the complicated mixtures of materials that we find in nature. Pure chemical elements are isolated from mixtures or compounds. Relatively frictionless environments are produced. Objects are dealt with at ultra-high or ultra-low temperatures to bring out certain characteristics. Pure-bred strains of mice, bacteria, or flies, often unable to survive in the natural environment, are developed for the purposes of experiment.

On the postmodern STS view, the "universality" of science is a product of the reproduction of these "local" special conditions of the laboratory in various places and transportation of the local results. Standardization of weights and measures, the shipping of purified substances, or of instructions on techniques for purifying substances help in transporting local conditions. The training of observers in laboratory techniques of measurement and observation allows the local observations to "travel" (Latour, 1987, chapter 6). According to Shapin and Schaffer (1985), Robert Boyle made experimental knowledge non-local by developing social networks of "virtual witnesses." Trusted individuals (gentlemen in Boyle's day, technical professionals in our own day) were to witness the experiments and their replication (Shapin, 1994).

Against this postmodern STS account is the claim of the usual view of scientific laws that they apply throughout the universe. According to this view the processes described by science are not simply local phenomena but processes that occur throughout the universe, although usually involved with other processes and factors. Furthermore, many processes or phenomena observed in biology, geology, and astronomy, such as huge and distant cosmological events or long-term geological or biological evolutionary processes, are not entities that exist within laboratories. The skepticism one sometimes encounters among scientists concerning large-scale cosmological speculations (Bergmann, 1974) and macro-evolutionary processes (long-term evolution of species and classifications above the species level) is due to there not being direct results of laboratory observations. During the first two-thirds of the twentieth century physical theories of the universe as a whole (cosmology) were considered speculative and less respectable than experimental particle physics. It was only when the theories of particle physics (grand unified theory, GUT, and later developments) became closely tied with cosmological results that the latter gained higher empirical status. A common objection to evolutionary theory by opponents of evolution is that large-scale evolutionary processes are not directly observed. Smaller-scale genetic changes over relatively short time periods are observed in the laboratory, but not the multimillion-year changes and trends of evolution. Both of these cases show how observational sciences that cannot be completely turned into experimental sciences can be considered somehow "less scientific" than experimental science.

The issue of scientific realism versus anti-realism becomes involved in this debate (see chapter 1). According to scientific realism, the theoretical entities discussed in scientific theories, such as atoms, subatomic particles, and fields, refer to real, although not directly observed, objects. Anti-realism claims that these theoretical entities are not real. According to the form of anti-realism known as instrumentalism, the language of theories is simply a tool for making predictions. According to fictionalism, this language of theories should not be taken literally. For the anti-realist, what is real is what is directly observed. This issue arises again with the treatment of science as local knowledge of experimental effects.

Should we believe that the purified and "cleaned up" processes observed in the physical laboratory actually occur (albeit mixed with and affected by many other processes) in the rest of the universe? The positivists and British empiricists (see chapter 1) wished to tie science tightly to sense-observations and demanded that theory be justified by being directly tied to sense obser-vations or else be treated as a purely symbolic instrument. In like manner, the technology and science studies approach implicitly demands that science be directly and tightly tied to laboratory demonstrations. Popper, Kuhn, and numerous other philosophers of science have made criticisms of inductivism as an account of the method of science. STS advocates have accepted many of these criticisms, at least of formal theories of induction, as an account of science. Yet, ironically, the model of delocalizing and extending the local knowledge of the laboratory as a model of the making of science (à la Latour and Hacking) is itself a form of inductivism. If so, then Hume's problem of induction still needs to be solved.

Local Knowledge and Technology Transfer

Although the debate over whether science itself is local knowledge may seem abstract and theoretical, it has importance for the fixing of the status of science in relation to indigenous knowledge. During the past few centuries, scientific knowledge has been seen as superior to indigenous knowledge. Western missionaries saw themselves as bringing the true religion to be-nighted savages. Similarly, colonial administrators saw themselves as bring-ing genuine knowledge of technology to replace superstition and "primitive" ignorance. In policy and practice indigenous knowledge has often been re-jected as "old wives' tales" or "gossip" (note the derogatory gender connota-tions), to be replaced by universal scientific knowledge. Colonial powers and Western scientific advisors have often ignored or discounted the tradi-tional knowledge of local peoples they were ruling or advising. Often the military might that allowed Western colonial powers to conquer and subjug-ate non-Western nations was seen as proof, via Western military technology, of the superiority of Western knowledge and peoples (Adas, 1989).

However, if Western science is a kind of local knowledge, then Western science and indigenous knowledge are put on a par, not in terms of political and military power, but in terms of claims to knowledge. Both are local knowledge systems to be evaluated on their own merits, especially with respect to applicability to local conditions.

Particularly in the cases of medicine and agriculture, both of which involve biological and environmental complexity, the strengths of local knowledge are evident. Often local farmers have detailed knowledge of the environment and its soils, weeds, and pests that scientific agricultural experts from the city or from other countries lack. Traditional Botswanan herding and Zulus in Mozambique had kept away the tsetse fly by burning the grazing lands and discouraging the growth of bushes. Colonial agriculture introduced rinderpest, killing livestock but allowing brush to regrow and bringing back the fly. One expensive American soil survey in Ghana was completed before the "experts" found that local farmers had already known and classified the main soil types just as the experts had (Pacey, 1990, pp. 190–1). Zapotec Mexican indigenous classifications of soil correspond closely to Western scientific classifications (González, 2001, p. 131). Traditional Zapotec farmers sometimes comment on the lack of knowledge of local conditions and successful techniques of visiting urban agricultural advisors (González, 2001, p. 221). Local intercropping methods turn out to increase yields when combined with chemical fertilizers over monocrop techniques with the same fertilizers (González, 2001, p. 170). On the Island of Bali, Dutch colonial agricultural administrators dismissed the complex pattern of religiously designated seasons of field use, abandonment, and rotation as a superstition, only to learn later that it was more efficient than their own continuous cultivation techniques. These examples suggest that because successful agriculture depends on complicated details of the local environment, local and traditional knowledge is sometimes more accurate and successful than applications of general scientific principles and technological techniques that lack knowledge of the details of the context of application. Many writers concerning agriculture and medicine in the less developed world have become more sensitive to the issues of local contexts of application.

Theories of the Differences between Technology and Magic

Despite the successes of local knowledge in indigenous technology in areas such as agriculture and medicine, the mixture of technology with magic in traditional societies is problematic. Many dismiss indigenous technology as "mere magic," or, at best, believe that Western science borrows indigenous knowledge, strips the magical dross from it, and extracts the pure gold of (Western) scientific knowledge. For instance, Western biotechnology firms collect plants known to cure diseases by local shamans in less developed nations, extract the chemicals believed to produce the effect, and patent the result, usually without benefit to the local communities that supplied the plants in the first place. This sort of behavior can be justified by the claim that the shamanistic knowledge was "merely magical," in contrast to the genuine knowledge produced by Western science and technology.

There have been several accounts of the relations of technology and magic in these societies, none of which has received universal assent. The opposing views have different implications for the status of the magical component of indigenous technology. The issue is still puzzling.

One view of the relation of technology and magic is that magic is simply a technology that does not work. The mathematician René Thom once quipped, "Geometry is successful magic," though he added, "is not all magic, to the extent that it successful, geometry?" (Thom, 1972, p. 11 n4). Magic, like technology, attempts to manipulate and control the world, both inanimate and animate.

The early anthropologist Sir James Frazer (1854–1941) claimed in his massive *The Golden Bough* (1890) that the core of magic is identical with science in its belief in the uniformity of nature (Tambiah, 1990, p. 52).

A common view of the decline of magic is that successful science led to it, as magic was increasingly considered less valid and useful than science (Thomas, 1971). A problem with this view is the fact that the witch craze, involving the belief in and persecution of witches, reached its height precisely when early modern physics was formed in the late sixteenth to midseventeenth centuries. One member of and leading propagandist for the Royal Society, the scientific society of England, Joseph Glanville, wrote a book on the reality of witchcraft. Only one minor member of the Royal Society disputed this work, and he was ridiculed for his beliefs – as well as his short stature. William Whiston, Newton's disciple and successor to Newton's chair at Cambridge, claimed that there was more certainty of the existence of witches than of Newton's gravity or Boyle's gas pressure (Webster, 1982, p. 98).

Another objection to the claim that science simply replaced magic is that magic was part of the worldview of a number of early modern scientists and contributors to the modern worldview. These included Paracelsus, a pioneer of chemical medicine, itinerant memory expert and magician Giordano Bruno, who first claimed that the sun is a star, and who propagated the view that the universe is infinite, and Newton himself, who believed he had merely rediscovered the law of gravity known to Pythagoras and other ancient sages, including the Egyptian God Toth or Hermes Trismegistus (Yates, 1964, 1968, 1972; Dusek, 1999, chapter 6). Leibniz, the inventor of symbolic logic and the differential calculus, believed that the binary number system had been invented by the ancient Chinese sage Fu Hsi (Dusek, 1999, chapter 11).

A third, and more obvious, problem for the claim that science replaces magic is that large numbers of people in the technological societies have quasi-magical beliefs of various sorts that survived the growth of modern science and technology. Indeed, the so-called New Age thought of the past few decades has involved for many educated people a revival of the respectability of magic. In Britain during the second half of the nineteenth century, a high water mark of faith in technological and scientific progress also saw widespread occultist movements (Turschwell, 2001; Bown et al., 2004; Owen, 2004).

It would seem that the rise of early modern science, far from discouraging magical thinking, coincided with its apogee in the witch craze. Furthermore, the dominance of modern science and technology in the late nineteenth and twentieth centuries coincided with a revival of occultism reminiscent of nothing so much as the superstitious "failure of nerve" that Gilbert Murray (1925) castigated in writing about the late Roman Empire. Historically, the thesis of the replacement of magic by science is dubious, to say the least.

The anthropologist Bronislaw Malinowski (1884–1942) claimed that magic begins where technology ends (Malinowski, 1925; Tambiah, 1990, p. 72). That is, technology is used to manipulate that which can be manipulated, but magic is used in cases where control fails. A different view of magic was also held simultaneously by Malinowski, and consistently by the philosopher Ludwig Wittgenstein (1889–1951). It holds that magic differs totally from technology in that magic is an **expressive** activity. That is, magic aims primarily not at practical external effects, but at the expression of internal emotions. Magic does not involve false factual beliefs, but is an expression of emotions such as anger in revenge magic or erotic passion in love magic. Magical language is rhetoric meant to sway the emotions, not describe reality. Wittgenstein's (1931) points appear in his highly critical notes on Frazer's *Golden Bough* (1890).

These competing views are relevant to philosophy of technology, insofar as Malinowski's main view makes magic a kind of illusory or mistaken attempt at an **extension of technology**, while Wittgenstein's view makes magic something non-factual, non-scientific, and not related to the manipulation of the physical world, quite different from physical technology.

Only insofar as magic is considered a kind of "**psychological technology**" does Wittgenstein's expressive view make magic a kind of (often successful) technology. The comprehensive systems definition of technology, which includes motivating aspects of culture, makes magic a part of technology, as technology now encompasses "all tools and culture" (Jarvie, 1967). Thus, differing views of magic construe the contrast between magic and technology in different ways. Is magic an ineffective attempt at technology? Is magic a non-technological form of expression more akin to art than technology? Or is magic a kind of psychological technology only *appearing* to be a physical technology?

Magical versus Technological Thought: Mythic versus Logical Mentalities?

Another area of controversy over the past century has been concerning the kind of reasoning involved in magic and technology. This can involve contrasting the "logic" of magic with that of technology. It also involves the issue of whether magical thought involves a kind of mythic structure quite different from technological reasoning. This contrast involves the logical reasoning involved in successful technology, either as applied science in some sense (see chapter 2), or as practical, means–end thinking or instrumental reasoning (see chapter 4).

Lucien Lévy-Bruhl (1857–1939), a French philosopher who dealt with topics that would now be considered to be in the area of social anthropology as well, presented in the early twentieth century a view that contrasted **"primitive" thought** with rational thought (Lévy-Bruhl, 1910).

The German Kantian philosopher of "symbolic forms," Ernst Cassirer (1874–1945), accepted and developed the same contrast, calling magic thought **mythic thought** (Cassirer, 1923), and further distinguishing between Aristotelian common sense and modern, formal, scientific thought. Cassirer's scheme is progressive, like that of Comte's three stages (see chapter 3). Recall that Comte's three stages are the theological, the metaphysical, and the scientific. Cassirer's three stages are: (a) mythic thought; (b) Aristotelian, common-sense thought, dealing with objects and qualities as fundamental, and based on subject–predicate grammar; and (c) scientific-functional thinking, based on mathematical and other functions. Cassirer analyzes the stages as different forms of the structuring of thought in terms of space, time, number, and causality, using Kant's philosophy of the organization and unification of perceptual knowledge through concepts or categories (Cassirer, 1923). (While at Yale University, Cassirer and Malinowski used to talk at length over lunch about "functional" thinking and its superiority to other forms of thought, but Cassirer meant mathematical functions, while

Malinowski meant social functions in terms of which a group is organized to serve ends, in so-called functionalist anthropology.)

Lévy-Bruhl claimed that "primitive" thought has a logic different from that of standard formal logic and the so-called **laws of thought** of traditional logic. The "laws of thought" are: (a) non-contradiction, i.e. something cannot be both A and non-A at the same time in the same respect; (b) excluded middle, i.e. anything is either A or non-A; and (c) identity A equals A. "Primitive" thought identifies opposites and rejects the prohibitions of contradictions of formal logic. It also disobeys the law of identity, as when a believer in totemism identifies himself or herself with the totem object: "I am a parakeet." It further identifies the part with the whole (as in magic in which harmful actions performed on hair or fingernails taken from the person to be affected supposedly act on that whole person). "Primitive thought" according to Lévy-Bruhl has a different sort of causality from causality as understood in civilized societies. Things are caused not through chains of secondary causes, but through direct action of non-located spiritual beings. Also, primitive space is not organized like the space of geometry.

Lévy-Bruhl, contrary to the accusations of some of his critics, did not claim that indigenous peoples were incapable of logical or causal thinking. He claimed that the group structures and solidarity of pre-state societies led to "collective representations" and a priority of the group to the individual. Lévy-Bruhl's critics have pointed out that the technology of non-literate or non-civilized societies exhibits the same practical task-oriented problem-solving structure as modern technology. Another, more recent, line of criticism of Lévy-Bruhl notes that in contemporary industrialized societies there remains much "primitive" or "mythic" thought in popular astrology, New Age mysticism and magic, and, less obviously, in advertising and political propaganda. Cassirer noted how similar he found Nazi propaganda to that of old magical and occultist writings of five centuries ago (Cassirer, 1948), while Malinowski noted the magical thinking involved in advertisers' attempt to link the purchase of automobile or toothpaste brands with the improvement of one's love life. Marshall Sahlins shows how implicit rules distinguishing pets from animals that can be eaten are as arbitrary as the taboos of non-Western societies. Eating of dogs in some East Asian countries horrifies Westerners who see nothing wrong with slaughtering cows for food, something that the Hindus of India find abhorrent (Sahlins, 1976). The authors of *Cosmodolphins* (discussed in chapter 9) analyze contemporary astrological beliefs in Russia and among New Age devotees using Cassirer's treatment of mythic thought (Bryld and Lyyke, 2000).

The anthropologist Evans Pritchard (1902–72) in early criticisms of Lévy-Bruhl argued that it is unfair to describe Western thought solely in terms of the reasoning of logicians and scientists and to ignore much popular culture and religion in the West, while describing non-Western cultures solely in terms of their magical and religious beliefs rather than their practical technology and agriculture (Douglas, 1980, pp. 27–30).

It is notable also that the public or popular image of technology in industrial societies often has elements of the magical, particularly with the development of atomic power. One member of the European Center for Nuclear research has spoken of "Scientists as Magicians: Since 1945" (Kowarski, 1971). Similarly, the theorists of nuclear weapon strategy have been called "the wizards of Armageddon" (Kaplan, 1983). Lévy-Bruhl himself in his last writings granted that both "primitive" and "logical" thought are present in all societies, although he claimed that the non-logical "primitive" thought was more obvious in non-literate, pre-state societies (Lévy-Bruhl, 1949).

Much of the contemporary rejection of Lévy-Bruhl is due to the appearance of racism in the contrast of "primitive" and "civilized" thought. Indeed, the original title of *Primitive Mentality* translates as *The Mental Functions of Lower [Inférieures] Societies*. Lévy-Bruhl denied that "savages" were incapable of logic and claimed that the alternative system of reasoning was quite as complex and organized as Western logic. Nevertheless, the chronicling of differences such as he presented have been used to argue for the inferiority and subjugation of persons of color in general and Black Africans in particular. However, even if the early Lévy-Bruhl was mistaken to associate "primitive" thought solely with non-Western, pre-literate societies, the distinction between "primitive" or Cassirer's "mythic" thought and logical thought holds.

Some of the critics have gone so far in attempting to draw similarities between the thinking of Western technological societies and pre-state, non-literate societies as to misrepresent the latter. Robin Horton argues that African traditional thought has the same hypothetico-deductive logic of laws, deduction, and testing that logical empiricist Carl Hempel or critical rationalist Karl Popper present as the logic of science (Horton, 1967). However, post-positivist philosophers of science have questioned whether this austere and rigorous logic really describes actual practice in the physical sciences. Furthermore, Horton grants that African traditional thought is closed and impervious to criticism or logical refutation. Evans Pritchard emphasizes the extent to which belief in witchcraft among the Azande is adjustable to all objections. This would suggest more the logic of Duhem, Quine, or Kuhn's account of science, with the use of the strategy of adjustment of auxiliary hypotheses taken to its extreme. Paul Feyerabend, the defender of "epistemological anarchism" and "anything goes" as scientific method, who became favorable to non-Western medicine when a traditional Chinese healer cured him of a urinary disorder that years of visits to various Western doctors had failed to cure, writes against his mentor, Popper, that he erred in thinking that:

> the suggested conceptual dichotomy (between empirical and metaphysical statements) corresponds to a real separation between statements that are part of scientific traditions, and . . . that traditions that do not know the separation or contain it only in a blurred form are improved by eliminating the blur . . . a closer look at the growth of rationalism in Greece, at the comparative effectiveness of "scientific" medicine and various folk medicines, at the comparative effectiveness, in special situations, of oracles and a "rational discussion" (described vividly by Evans Pritchard) shows that this assumption too is highly questionable. (Feyerabend, 1981, pp. 21–2)

(On Popper's demarcation of science and non-science, see chapter 1.)

Separating the Technology from the Magic in an Activity

Even if one accepts a distinction between magic and technology in non-literate, non-state societies, there remains the issue of distinguishing magic from technology when they function in the same activity, such as fishing or agriculture. Many practical activities in indigenous societies involve both practical technological activity and magical chants or the casting of spells. This is most evident in pre-literate, pre-state societies, but is not absent in contemporary society. Workers in dangerous occupations, such as sailors and miners, often have superstitious rituals, even if not as elaborate as those of indigenous peoples. Boat-building, weapon-making, fishing, planting, and harvest have their appropriate ceremonies and rituals. To the people involved both what we would see as technological and what we would see as magical actions are essentially involved in the task.

A traditional African or Japanese sword-maker engages in heating, hammering, and quenching, accompanied by various mythic rites (Ellul, 1954, p. 20; Bronowski, 1973). The Western technologist would claim that the mythic prayers and ceremonies could be dispensed with, leaving the pure metallurgical process. However, anthropologists such as Malinowski and later functionalists emphasize that the magical rituals often function to unify and motivate the work group of fishers or farmers (Malinowski, 1922). The scientifically influenced observer of these activities may wish to separate the effective technological activity, the part that "works," from the magical components. One Polynesian twentieth-century navigator courageously tried this, and successfully performed traditional ocean canoe navigation without the magical rituals.

One of the peculiar results of the inclusion in technology of prayers and rituals as "psychological technology" is that the magical ceremonies themselves can be analyzed as technology, now a psychological technology of motivation rather than a physical or biological technology of hunting or fishing. One problem with this view of rituals and chants as "psychological technology" is that it so broadens the notion of technology as to make it indistinguishable from other spheres of life. However, given that technology is so intertwined with and so penetrates all aspects of human life, devotees of technology studies might not see this as a problem. Even the Popperian anthropologist-philosopher Ian Jarvie includes psychological technology within technology, and concludes one examination by admitting that his notion of technology expands to include all of culture and society (Jarvie, 1967, p. 61).

However, one can be led to a different view of the role of rituals and chants if one accepts the technological system definition of technology, which includes social organization as a part of technology, or Pacey's "technology practice," which includes culture as well as society as a part of technology. This view could allow the ritual aspects of indigenous technology, in their role of motivating or timing the rhythms of work, to be part of technology. The line of demarcation between magic and technology is dependent on one's views on the status of so-called psychological technology as technology, the understanding of technology as applied science, and the definitions of technology as hardware or tools, versus the definition of technological system as including social organization and the relevant cultural motivations.

Traditional Chinese Technology and Science and the Possibility of Alternative High Technologies

The technologies and science of traditional India, China, and Japan are examples not of ethnosciences, but of the technology and science of highly cultured civilizations that did not follow the route of the ancient Greeks or Renaissance Italy. Do these sciences and technologies show alternative possibilities, "what might have been," as alternative routes to a high technology that were aborted by the dominance of Western science? Many would say no.

The usual view has been that these cultures did not have science. Einstein (less informed here in history and cultural studies than in physics and mathematics) often puzzled over why China never developed science. Even in the 1990s a blurb for a popular book on scientific method by a nuclear physicist and educational theorist, Alan Cromer, states that "he nails his thesis against the doors of . . . political correctness and multiculturalism reiterating his view that the core of scientific thinking is a uniquely Western discovery" (Oxford University Press, 1995–6). Cromer claims that "the Chinese, seeing everything from their own perspective, learned nothing about the outside world." It is, then, odd that a 1315 Chinese map was the basis for a Korean map of 1401 that is more complete than any map made by Europeans at that time. Cromer also claims that (presumably unlike himself) "the Chinese didn't develop objectivity" (Cromer, 1993). By the time Cromer wrote this, a score of volumes by the biochemical embryologist and science historian Joseph Needham had long buried this ignorant view. Starting in the late 1940s, Needham uncovered for the West the vast riches of traditional Chinese science and technology in a project that some have called the greatest historical achievement of the twentieth century.

The traditional Chinese not only had a technology far ahead of that of the West up until the Renaissance, but many Western technological devices seem to have migrated from China to Europe. Francis Bacon, in arguing that the Greeks and Romans did not think of everything, cited what he believed were three modern Western inventions unknown to the ancients: the magnetic compass, printing, and gunpowder. Unbeknownst to Bacon, these inventions were, though indeed unknown to the ancient Greeks, products of medieval China that had traveled to the West.

Not only did numerous Chinese inventions travel the Silk Road through Central Asia to the West during the Middle Ages (stirrups, wheelbarrow, fishing rod with reel, and many others), but also in some areas Chinese technology was ahead of the West until the twentieth century, particularly in "holistic" areas of medicine, ecology, and earth science. The Chinese had maps of shifting magnetic field lines and declination on the surface of the earth centuries before Westerners learned from them about the magnetic compass (Dusek, 1999, chapter 2). The Chinese had hormone therapy, both thyroid hormone and estrogen therapy, in medieval times. They had biological pest control (the use of insects eating insects rather than chemical pesticides) centuries before American agricultural advisors in the nineteenth century brought these ideas to the West. The ancient Chinese had seismographic observation of earthquakes by Chang Heng as early as 100 CE (Ronan, 1978, volume 2, pp. 300–5).

There was deep drilling for brine as early as 100 BCE, with traditional wells in the modern period extending to at least 4800 feet (Needham, 1954, volume 1, p. 244; Vogel, 1993, p. 86). There were deep natural gas wells penetrating a thousand feet into the ground many centuries before the West achieved such things. Chinese drilling techniques percolated to the West through Dutch visitors beginning in the seventeenth century and culminating in detailed descriptions by French missionaries in 1828. French engineers soon copied the Chinese techniques. In fact, the famous 1859 Oil Creek well in Pennsylvania of Colonel E. L. Drake used techniques probably borrowed from China, either via the French or from Chinese-American indentured Chinese railroad workers (Temple, 1986, p. 54). Thus, the American oil and gas giants that only now are moving into the "Stans" of Central Asia after the Afghan War use drilling techniques that Chinese used over 1000 years ago and that the USA borrowed from them.

The superiority of Chinese over Western technology before the Western scientific and industrial revolutions is widely granted by contemporary historians of technology. More controversial are Needham's claims about Chinese science. Needham himself, goaded by some visiting Chinese graduate students, was initially puzzled why China did not develop physical science. If one means by science the combination of mathematical laws of nature and controlled experiment that began in the sixteenth century with Galileo, then China never developed science.

Needham, following Edgar Zilsel, suggested that the European notion of universally applicable laws in jurisprudence was transferred by figures such as Bodin and Kepler in the sixteenth and seventeenth centuries to laws of physical nature. The notion of laws of society was transferred in Europe to the notion of a Divine Lawgiver that legislated for the physical universe. Needham traces an intermediate stage in the late Middle Ages, when a rooster that laid an egg as well as a three-legged hen were put on trial and executed as "unnatural." He concludes that only a society crazy enough to have a trial for an egg-laying rooster could have the fanatical faith in the order of nature that yielded a Kepler or a Newton despite nature's chaotic appearances and the inaccuracy of then available observational data (Ronan, 1978, volume 1, pp. 301–2). A talk once summarized the situation with a title that must have puzzled those not yet familiar with Needham's opus:

"What Do You Do when a Rooster Lays an Egg? Or the Social Origins of Modern Science" (Walter, 1985).

China had a high quality of purely observational science in astronomy, geology, and biology. The Chinese, uninfluenced by the Aristotelian notion of the separateness of the pure "quintessence" or aether that made up the stars and planets from the grubby earth and water that made up our earthly abode, observed the birth of new stars (novae) and sunspots centuries before did Tycho Brahe and Galileo. In the West, the birth of new stars and the existence of sun spots were initially rejected by the Jesuit scholastics as inconsistent with Aristotle and St Thomas Aquinas' view of the heavens. Ironically, when Jesuits Matteo Ricci and others brought Western geometry, mechanical clocks, and astronomy to China in the early sixteenth century, the church back in Rome was banning the views of Galileo and put him under house arrest (Sivin, 1973; Spence, 1984). Thus after a few initial years, the theories of Copernicus, Galileo, and Kepler about the sun-centered solar system and motion of the earth could not be taught openly and directly to the Chinese by the Jesuits. Ironically, the heathen Chinese were told by the missionaries that their infinite universe, in which stars floated without support in empty space and in which stars were born and died, was primitive superstition, just as those very views were spreading in Europe.

Despite the high quality of Chinese observational astronomy, Chinese computational astronomy had degenerated, even though in ancient times it had been approximately equal to that of the Babylonians and early Greeks. Furthermore, the Chinese never developed physical and astronomical laws of motion, as did the West. Only one school of thought in ancient China, the Mohists, who were lower-class artisans and military engineers, developed quantitative ideas of optics and mechanics that might have developed into something like Western physics (Graham, 1978). However, the Mohists, as artisans with independent power as mercenaries, were suppressed with the unification of the Chinese Empire around 200 BCE. Most Chinese physical thought emphasized the inexactness and indeterminacy of physical measurements (Sivin, 1973).

At a purely qualitative level, the Chinese view of the universe as an evolving process-natured entity, with indeterminacy and lack of permanent underlying substances, looks more like twentieth-century science than like the classical mechanical view of the seventeenth- and eighteenth-century West (Dusek, 1999, chapter 3). Needham even suggests that the Chinese were, in effect, attempting to leap to a worldview akin to that of relativity and quantum theory without going through the intermediate stage of classical, deterministic mechanics, and Newton's absolute space and time (Ronan, 1978, volume 1, p. 292).

When Chinese gunpowder in cannons carried on ships guided by the Chinese-invented rudder and compass bombarded and defeated the Chinese troops to enforce the opium trade on the resistant Chinese monarchy, Chinese science declined. Just as the Chinese were denounced as degenerate opium addicts for imbibing the drug that the British forced them to accept, so their science was denounced as primitive and superstitious.

The decline of science occurred in many parts of the world just before, during, and after the European invasion and domination. The British Raj in India prohibited the teaching of indigenous Indian mathematics to the Indians. This suppressed memory of the mathematical tradition of Kerala, which in the fourteenth to sixteenth centuries developed techniques similar to the differential and integral calculus invented shortly afterwards by Newton in Britain and Leibniz in Germany (Joseph, 1991, pp. 299–300).

The West suppressed records of older, indigenous engineering achievements after it came to dominate. British colonial rulers prohibited teaching Black children about the African stone city of Great Zimbabwe in Southern Rhodesia. Meanwhile, the city was progressively looted and defaced and looted by the colonizers. Rhodesian Blacks were taught that African Blacks had never built stone buildings or had "civilization," in the sense of cities. An old colonial tradition, stemming from the German explorer Mauch, attributes Zimbabwe to Sheba of the Bible, in order to claim that it not have been built by African Blacks. Another tradition, alive even today, associates Zimbabwe or nearby Angola with Ophir in the Bible, associated with Phoenicians rather than Blacks, despite archeological evidence that it was built by Africans, not by visitors from the Middle East (Davidson, 1959, pp. 250–64; Kinder and Hilgemann, 1964, p. 221). The fact that the earliest mineral mines had developed in prehistoric Black Africa was likewise long ignored.

If European military technology and strategy had not overwhelmed indigenous populations on other continents, could alternative sciences and science-related technologies have developed? Technological determinists (as presented, for instance, in Robert Heilbronner's arguments) tend to claim that there is one predetermined route for science, and, since modern technology is identified with applied science, there is but one route to high technology (see chapter 6). Could an alternative, more holistic, ecological, and medical technology have developed in an isolated China to levels reached by Western technology?

Perhaps Chinese decline would have prevented it in any case. Just as Vasco de Gama, Columbus, and Magellan were embarking on their world girdling and eventually dominating voyages, the Chinese pulled back their mighty fleets, which under Admiral Cheng Ho, the "Three Jeweled Eunuch," had sailed to and traded with Africa (Ronan, 1978, volume 3, pp. 123–35). Chinese magnetic and geographical lore, during this period far ahead of that of the West, was still preserved in the geomancer's compass, but progressively overlain with superstition.

Likewise, Arabic science and technology had reached heights far above the medieval West, and were the source of the revival of Western science and logic, as Arabic texts were translated into Latin in the "renaissance of the twelfth century" CE. There were factories in the Arab Middle East as late as the seventeenth century, but there too decline soon set in (Rodinson, 1974). Some World Systems theorists in economics, in their most radical phase, suggested that the decline in the non-European regions was in fact a product of the expansion of Western economic penetration (Frank, 1967).

Yet the simultaneity of general Asian decline and Western rise to power may be a coincidence. The decline started in China, in the Mughal Empire of India, and in the Middle East shortly *before* the Western penetration. The case of Cheng Ho's fleet being disbanded in China is a case in point. Could there be economic effects of the encroaching European system before the actual explorers and armies arrived? Certainly in the Americas there is a case for such a cause of decline preceding the arrival of Western settlers. It appears that a very few European explorers on the east coast of North America were sufficient to spread major epidemics into the center of the continent decades or even centuries before large numbers of European settlers arrived there. The triumph of the armies of the Spanish conquistadors in Mexico and Peru over Native American armies that vastly outnumbered them, especially in the case of a few dozen troops of Pissaro conquering the Inca Empire, has always been puzzling. The situation is easier to understand when one realizes that the vast majority of Aztecs and Incas were suffering from fever and dysentery spread from the earlier coastal landings of Europeans before the conquistadors arrived at the capital cities of the empires.

However, economic causes of disruption of the economy of the Mughal Empire in India or of the Ottoman Empire prior to the arrival of the European invaders, if such exist, are presently unknown, though various conflicting hypotheses are floated in the literature of economic history and world systems theory (see, for instance, the journal *Review*, passim, and symposia in *Modern Asian Studies*, 24(4), and the *Journal of Economic History*, 29(1)). It may be that trade with the West prior to the European invasions "softened up" and undermined the Middle Eastern and South and East Asian economies for European domination and colonization.

Gunder Frank, in his more recent work on "re-Orientation," has suggested that the dominance of the West and the submission of China may be a relatively temporary (two-century) accident, due to an economic crisis in the silver trade. According to Frank, China will have moved back to its rightful, millennia-long, and traditional self-designated place as "Middle Kingdom" or center of the world by the mid-twenty-first century (Frank, 1998). One suggestion is that the use by the West of the precious metals looted in Mexico and Peru supplied the West with a means of trade not supplied by its paltry manufactures, and the drain of silver from China because of the British introduction of opium and the resulting widespread addiction may have shifted the world balance of trade.

In contrast to the technological determinists, social constructionists claim that the appearance of the necessity and inevitability of our present scientific theories and "facts" is a product of social processes of consensus and knowledge dissemination. A variety of interest groups and factions contribute to the development and stabilization of the science and technology that we have (see chapter 12). Could the same contingency be true of our present science and technology as a whole? Needham's work on China, as well as similar work on India, Latin America, the Middle East, and Africa, raises this question.

The answer, whichever it is, has major political implications. If an alternative, perhaps more ecologically sensitive and holistic, science (such as the indigenous sciences of China, India, and other non-Western cultures) is possible, then the claims that our present science and technology is a product of its own inexorable logic are false. Both the pessimists, such as Ellul, and optimists, such as the modern technocrats, may be rationalizing present arrangements. Both culture-critical pessimism and the technocratic ideology criticized by critical theorist Habermas and others (see chapter 4) may obscure genuine alternatives. On the other hand, if the theses of autonomous technology and technological determinism, and the claim that Western science and rationality are truly universal, are correct, then the hopes for such an alternative science and technology held by feminists, deep ecologists, and self-described radical reformers of science and technology are a dangerous illusion (see chapters 9 and 11, and the discussion of the "science wars" in chapter 1).

Study questions

1   Is it possible to sharply separate the purely technological aspect of a cultural activity from the ceremonial, religious, or other aspects of the activity?

2   How would you distinguish magic, science, and religion?

3   Is scientific knowledge a truly universal description of nature in itself, or is it a kind of Western "local knowledge" that has spread over the globe through the building of Western-style laboratories as environments for that "local knowledge"?

4   Is indigenous knowledge (say of agriculture or medicine) merely a simplified or inaccurate example of scientific knowledge or is it a completely different sort of thing from scientific knowledge?

5   Do you think that China, if isolated from the West to the present, would have developed its own brand of advanced science and high technology?

11

Anti-technology: Romanticism, Luddism, and the Ecology Movement

> One impulse from a vernal wood
> May teach you more of man,
> Of moral evil and of good,
> Than all the sages can.
>
> Sweet is the lore which Nature brings.
> Our meddling intellect
> Mis-shapes the beauteous forms of things:
> We murder to dissect.
>   William Wordsworth, "The Tables Turned"
>
> Things are in the saddle and ride mankind.
> Ralph Waldo Emerson, "Ode Inscribed to W. H. Channing"

Some intellectual and social movements are highly optimistic about technology. They totally praise and are uncritical of it. Positivism (see chapter 1), orthodox Marxism, and technocracy (chapter 6) are examples. Other movements have been critical of technology and pessimistic concerning the present direction of technology. Romanticism is a movement that has propounded a number of main themes that have been taken up by many later anti-technology movements. Romanticism arose in the late eighteenth and early nineteenth centuries within poetry, philosophy, and visual art, and continued through much of the nineteenth century in music.

This chapter examines a number of anti-technology movements, including the Romantic Movement, the original Luddism of the machine-breakers at the beginning of the industrial revolution, modern accusations and self-designations of Luddism, and the late twentieth-century deep ecology and other radical ecology movements.

Romanticism

Romanticism arose, in part, as a reaction against the industrial revolution of the late eighteenth century. The pollution, urban poverty, and ugliness of the newly industrializing cities repelled many writers and thinkers. Romanticism was also a reaction against the worship of reason (see chapter 4) and denigration of the emotions that was promoted widely by writers of the Age of Reason and Enlightenment eras of the previous century.

The French philosopher and educational theorist Jean-Jacques Rousseau (1712–78) was the great forerunner of and inspiration for many of the romantics. Rousseau's earliest significant work, his prize essay *Discourse on the Sciences and the Arts* (1750), argued, contrary to the spirit of the age, that the development of civilization, science, and technology had been harmful to morals and society. Rousseau praised the archaic civilizations, warlike and heroic, and claimed that civilization led to weakness and decadence. In his educational treatise *Emile* (1762), Rousseau argued, as would many later romantics, that discipline and direction suppressed the natural impulses and creativity of the child. Rousseau even recommended that reading should not be taught until the child had reached the level that we should call middle school. Education should be pursued by communing with nature in the woods and fields, not in the classroom. The romantic cult of the child appears again in Wordsworth and many others through twentieth-century progressive education. The notion of the "noble savage," the pre-civilized person, superior in spirit and personality to the alienated products of civilization, also influenced much later anthropology and cultural theory.

An example of Rousseau's impact is the fact that Immanuel Kant, who in his later years kept to a schedule so rigid that townspeople could set their watches by his passing their houses on his daily walk, missed his walk only twice, once when news of the French revolution arrived, and the other when one of Rousseau's books arrived in the mail (Cassirer, 1963). Numerous Europeans wept and swooned to Rousseau's sentimental novel *The New Eloise* (1761). One noblewoman, dressing for a ball, happened to pick up a copy that had arrived. She bid her coachman wait a minute, while she read a bit, asked him to wait longer, and finally made him wait all night, never arriving at the ball.

Certainly the rise of science and its success, including the success of Bacon's inductive method (see chapter 1) and of the use of mathematics in formulating laws of nature by the early modern physicists and astronomers, stimulated and reinforced their emphasis on reason. The success of Galileo, Newton, and others in formulating laws of motion and mathematically predicting the motions of the planets led to the belief that the mathematical and experimental methods would solve all human problems, including those of ethics and politics. The passions, in contrast, were seen as unruly and in need of suppression or channeling by reason. The romantics opposed this and championed the importance of emotion and passion. The romantics downplayed reason in favor of passion and imagination or, in the case of the German romantic philosophers, praised reason but meant by it something that was highly intuitive and very different from the reason that the followers of the experimental or mathematical methods had previously praised. Romantic philosophers such as Schelling and other "nature philosophers" claimed that "transcendental reason," in the form of intuitive insight and a higher imagination, instead of observation and experiment, would explain the structure and ultimate reality of things (see chapter 4 on transcendental reason).

Romanticism also questioned the view of the world that the philosophers of early modern science had presented. Science, particularly physics, was successful in describing the world in terms of mathematics, of quantity. In addition, physics explained the world in terms of atoms, entities unobservable in direct perception. Physics was described in terms of mass, length, and time, not of other qualities. For instance, experienced colors, sounds, smells, tastes, the so-called "secondary qualities," were explained in terms of the spatial, "primary qualities." Secondary qualities were considered a subjective product of our sense organs and mind and not as real or fundamental as the primary qualities.

The romantics reacted against this claim by emphasizing that what is real is what we directly perceive in terms of colors and sounds. Nature, as directly experienced, is real, while the physicists' description of nature in terms of atoms and geometry is a lifeless abstraction. A twentieth-century philosopher and mathematician, A. N. Whitehead, sympathetic to the romantic critique, called the belief in the reality of abstractions and the unreality of what is directly perceived "the fallacy of misplaced concreteness." It confused our intellectual abstractions with reality. (See similar ideas in chapter 5.)

The denigration of direct sensory experience of qualities, of colors and sounds, and the elevation of quantity and number as the ultimate reality may have had consequences for the justification of lack of concern for the ugliness, filth, and pollution of the new industrial cities. Both physics and economics dealt with quantity. Physics described reality and economics calculated profit and loss. Concern with matters of beauty and ugliness was rejected as trivial and irrelevant. Writers after romanticism, such as Thomas Carlyle (1795–1881) and Charles Dickens (1812–70), pointed out how the quantitative approach to all things, including morality, had deadened people's taste and sympathies and made education desiccated and boring (Carlyle, a conservative, coined the term "cash nexus," which was borrowed by Marx and Engels in *The Communist Manifesto*).

One feature of much romantic thought that has been taken up by ecological critics of contemporary science and technology is holism. Holism is the claim that the whole is more than the sum of its parts. That is, the whole system has qualities and characteristics that are not those of its parts.

Romantics opposed the analytical and atomistic approach. The British poet Wordsworth famously said "They murder to dissect." Here in a nutshell is the romantic distrust of analysis into parts of atoms and the belief that such an approach destroys what is worthwhile in the organism or system being studied. William Blake similarly denounced atomism and the approach of John Locke:

> The atoms of Democritus
> And Newton's particles of light
> Are sands upon the Red Sea Shore
> Where Israel's tents do shine so bright.
> William Blake, "Mock On"

For Blake, to teach atomistic doctrines was "to educate a fool how to build a universe of farthing balls" (Bronowski, 1965, p. 137). Like some contemporary postmodern science critics, Blake associated atomistic doctrines and inductivism with the political establishment that he opposed.

Romantics did not reject science as such, but thought that a science different from the mechanical approach was needed. In contrast to the mechanical view, romantic physicists presented a "dynamical" view that emphasized forces rather than objects. Some historians of science have suggested that the field approach to physics of Oersted and Faraday, as well as the notion of the conservation of energy, owed part or all of its formulation to romantic conceptions of nature as unified (Williams, 1964). Romantic nature philosophers wanted an intuitive rather than an analytical approach to nature.

For the romantics Nature is what is directly perceived in terms of qualities, the nature grasped in ordinary perception, the nature portrayed by the artist. Also, "Nature" is valued over civilization and refinement. Being natural

Box 11.1

Holism

Often holism claims that the whole determines the features of its parts. The term was coined by the early twentieth-century South African premier and military leader Jan Smuts (1870–1950), who used to read Kant for relaxation in his field headquarters tent (Smuts, 1926). Opposed to holism is atomism, which analyzes wholes into their smallest components. Also opposed to holism is reductionism, the claim that the parts are smaller components, which fully explain the whole or system, and/or that they are more real than the whole system that they compose. To use a modern example, a reductionist biochemist or molecular biologist would analyze a living organism into its atoms and molecules. A holist organismic biologist would focus on the functions and behavior of the whole organism, and deny that some of these are adequately accounted for in terms of the characteristics of the atomic parts. Writers such as organic chemist and ecology activist Barry Commoner (1967) criticize reductionism and defend holism, claiming that living organisms and natural systems must be understood as wholes. Many political ecologists are holists, emphasizing the interconnectedness of all things in the environment and claiming that the analytic, atomistic approach leads technologists to overlook the environmental side-effects of their projects. There are various degrees of holism that are often confused in discussions of the issue. The most extreme holism is **monism**, which claims that there is only one object (see box 11.2). A less extreme holism is **organicism**. In organicism, the system is a whole that determines its parts, but the parts have relatively independent existence. A number of twentieth-century biologists, such as J. B. Haldane and Paul Weiss, have been organicists, in opposition to mechanists as well as vitalists (who claim the existence of a separate life force). Another, yet weaker, form of holism is **relational holism**. This positions claims that all the elements of a system are significantly related to one another, and that an understanding of the relations is impossible simply from the elements that are related. Some logical positivists deny that relational holism is really a holism at all, and claim that the acceptance of the reality of relations allows one to defend atomism and mechanism (Bergmann, 1958). Process philosophers, on the other hand, claim that the relations are the only reality, and that the relata, if they exist at all, are further, lower level relations (see box 12.2). Similarly, some anti-atomist, anti-mechanist biologists, such as evolutionists Stephen Jay Gould (1941–2002) and Richard Lewontin, oppose holism because they identify it with monism, although they are close to the organicism and relational holist positions (see Dusek, 1999, chapter 1).

rather than artificial is valued. Jean-Jacques Rousseau, who praised "natural man" and despised the artificiality and hypocrisy of the French society of his day, was the great initiator of this attitude.

The romantics saw the industrial revolution and the new technology as destroying both nature and the human spirit. The belching smokestacks and polluted streams and rivers of the industrial centers destroyed nature, while the crowded, unhealthy living conditions, the repetitive work, the poverty of the workers, and the greedy pursuit of wealth by the owners destroyed human character. Technology itself – the steam engine, the railroad, the mill – was frequently seen as the culprit. Blake famously wrote of the "dark satanic mills" in a poem ("And did those feet . . .") that became a popular hymn. Ruskin likened traveling by railroad car to being shipped like a package (Schivelbusch, 1979). Nature, in contrast to the artificiality of civilization and the ugliness of cities, was seen as a source of wisdom and inspiration. "One impulse from the vernal wood," claims Wordsworth in the poem at the head of this chapter, will deliver wisdom.

Later "back-to-nature" movements – such as the arts and crafts movement of nineteenth-century Britain, the German youth movement of the early twentieth century, the counterculture of the 1960s and after, and the ecology movement of the past few decades or the New Age movement – share romanticism's praise of wild nature, criticism of artificiality, and distrust of technology. William Morris's late nineteenth-century arts and crafts movement rejected the uniformity and lack of imagination of mass produced objects and emphasized a return to handicraft (Thompson, 1977). The German youth movement combined outdoor hiking and camping with a disdain for the "lifeless" abstractions of physics and technology (Heer, 1974) (Heisenberg was a follower of the youth movement; see chapter 6).

Luddites

During the past half century the term **Luddite** has most commonly been used to disparage opponents of technology. Anti-nuclear demonstrators, opponents of computerization, and other critics of technology have been dubbed Luddites by the defenders of the technology. Sometimes people in organizations are called Luddites, half jokingly, simply because they are slow or reluctant to learn to use new office technology or software. More recently, some members of the ecology movement and others who see all of modern technology as harmful have proudly called themselves neo-Luddites.

The original Luddites were weavers and other textile workers in Britain at the end of the eighteenth and beginning of the nineteenth centuries whose home handwork was being made obsolete by mechanized looms and weaving factories. The Luddites smashed the factory machinery in protest at and opposition to the new factory system. They claimed to be followers of "General Ned Ludd" as leader of the movement, who may have really existed or may be a mythical figure. Numerous letters and manifestos by different individuals and groups were issued in the name of General Ludd or "King Ludd." The original Luddites apparently were acting primarily from economic motives. They were being put out of work by the lowering of prices for cloth. They were being forced from their traditional craft-based homework into the factory system, subjected to labor discipline, and making less money in the process (Hobsbawm, 1962; Thompson, 1968).

The modern use of the term Luddite is somewhat misleading. Many of the modern anti-technology crusaders whom the supporters of technology disparage as Luddites or who proudly call themselves neo-Luddites are generally concerned not with direct impoverishment or job loss but with lifestyle issues (compare the mainly literary figures described in Fox (2002) as later members of the Luddite tradition). The so-called neo-Luddites are really more akin to the romantics in rejecting technology as alienating and inimical to a well lived life. (Some romantics, such as the British poets Byron and Shelley, defended the Luddites, but seem to have done so more from radical political views and sympathy for the downtrodden than from romantic nature philosophy.)

Chellis Glendinning, a psychologist, issued a "neo-Luddite manifesto" (1990). Glendinning is not solely concerned with lifestyle issues, but sees contemporary technology as a genuine threat to life and health. She had previously researched people who had suffered from various nuclear and chemical technologies, such as pesticides and medicines that have produced cancer and pain. As a psychologist, Glendinning compares our social addiction to technology to drug and alcohol addiction, considering the writers of techno-hype as enablers. Glendinning, like most other critics of contemporary technology, does not demand the elimination of all technology, but wants a development of different technologies more amenable to human well-being and political democracy than present technologies. Social scientists talk of technological choices versus social ends, claiming that the choices for technology development are made with an eye toward profit rather than with an eye toward improving the society.

Ecology, the Conservation Movement, and the Political Ecology Movement

The term "ecology," for a branch of biology dealing with the interrelations of the members of natural communities of organisms, was coined in 1866 by the German evolutionist Ernst Haeckel. Ecology in its American version in the first third of the twentieth century, borrowing ideas from the Dane Eugenius Warming and others, and developed notably by Frederic Clements, focused on the succession of plant and animal communities in a given area, as in the succession from pond to marsh to forest. The succession of plant life in a given area develops to a "climax" community. The community succession notion was one of progress toward a harmonious equilibrium. (Here we can see the influence of the notion of historical progress that dominated technocratic thought as well as many other philosophies of history in the nineteenth and early twentieth centuries.) This early ecology also treated plant and animal communities as organisms. Clements borrowed this approach from the pre-Darwinian evolutionist, philosopher, sociologist, and later social Darwinist Herbert Spencer. Both Spencer, with his organismic theory of both natural communities and human societies, and Haeckel, with an overall organismic view of human societies, show the strong tie of early ecology with holism.

There is a darker side to Spencer and Haeckel's influence, with the ideology of social Darwinism and imperialism common among many of the followers of Spencer, and with Haeckel's Monist Society, which according to some devolved into the Nazi movement after Haeckel's demise (Gasman, 1971). US President Theodore Roosevelt combined imperialist social Darwinism, advocating domination of the "savage" races by Anglo-Saxons as in the Spanish American War, with innovative support of conservation and the National Park system. Roosevelt's military adventures are celebrated in the murals that cover the walls of the vestibule of the American Museum of Natural

History. I, who once worked as a volunteer at the museum for a summer, and another biologist who frequently visited there, had never noticed these, despite entering the museum hundreds of times through that hall.

The Nazis had a strong interest in ecology and preservation of nature, combined with their racial extermination policies, understood as parts of a unified policy for biological health. Germany under the Nazis led other countries by decades in attempting to eliminate smoking as a cause of lung cancer (Proctor, 1999). The Nazis, with their encouragement of breeding the Aryan "master race," as well as their policy of eliminating Jews, Slavs, Gypsies, and gays, had a policy of fostering what Foucault (1976) called "biopower." There is a famous photo of Hitler petting a fawn, to show what a lover of animals he was. Amazingly, the concentration camp at Dachau had a health food herb garden, which fed guards, as well as some prisoners who were later exterminated (Harrington, 1996).

The tie of ecology to holism was revived among biologists and philosophers at Chicago and Harvard universities in the 1920s and during the 1970s in the political movement called the ecology or green movement. One of the late nineteenth-century ecologists, the Scotsman Patrick Geddes, applied his ecological approach to city planning and in turn influenced Lewis Mumford (Boardman, 1944) (see chapter 8 on Geddes and Mumford).

With the Midwest dust bowl of the 1930s and the rise of Franklin Delano Roosevelt's New Deal during the Great Depression, holistic ecological thinking based on the climax community notion came to dominate the conservation movement. Mechanized agriculture in the form of the tractor was claimed to have caused the dust bowl. The "sodbuster . . . had bound himself by a different set of chains, those of technological determinism" (Worster, 1977, p. 246). The holistic and organismic approach to land management was contrasted with the partial, fragmented, and atomic approaches of individual interests, such as those of the farmer, real estate developer, or timber company. Not since the Romantic Movement and the industrial revolution in the early nineteenth century was there so great a conflict as that in the 1930s between nature and society (Worster, 1977, p. 237 and chapter 12 passim).

Scientific ecology began to criticize the rigid succession and climax community notions that were beholden to the theory of historical progress and organismic equilibrium. A. G. Tansley of Oxford, although a follower of many of Clements's ideas, criticized the notion of a unique, natural climax. Later, Tansley began to criticize the organismic model of the living community. With the growth of energy flow conceptions of ecology based on the laws of thermodynamics (another notion vaguely adumbrated by Spencer in the previous century), late twentieth-century ecology, in contrast, has often emphasized process and disequilibrium.

The emphasis on disequilibrium grew gradually, as a Darwinian, competition-influenced model of equilibrium was also developed out of population science in a mathematical form. For instance, Edward O. Wilson and Robert MacArthur's *Theory of Island Biogeography* (1967) was formulated in conversations that began concerning the "balance of nature" that MacArthur reformulated mathematically as a theory of equilibrium (Quammen, 1996, p. 420).

Tansley's terminology of "**ecosystem**," rather than super-organism or community, was melded with Charles Elton's notions of food chains and energy flows. The competition-based Darwinian "economy of nature," with its version of the economic "invisible hand" leading to equilibrium and optimality, continued to have influence. However, the energy/flow economic model emphasized rather economic notions of consumption and production, input and output. The "new ecology" also emphasized planning and management of the environment as a whole, just as the technocrats emphasized the planning and management of society.

H. G. Wells, until near the end of his life a partisan of technocratic planning, and the evolutionist Julian Huxley wrote a chapter called "Life under control" in their elementary biology textbook in the 1930s (Worster, 1977, p. 314). H. T. Odum, one of the Odum brothers who did US Atomic Energy Commission funded work on the ecological effects of H-bomb tests, in his *Environment, Power and Society* (1970) expounded the technocrat's dream of a society constructed in a carefully manufactured pattern. Kenneth Watt demonstrates in his book *Ecology and Resource Management* (1968) that the new ecological principles lend themselves easily to the agronomic desire to "optimize harvest of useful tissue" (Bowler, 1992, p. 540). Watt's somewhat *ad hoc* systems theoretical approach does not find favor with the mathematical population biologists who are influenced by Marxism. Richard Lewontin and Richard Levins even went as far as to satirize and ridicule his systems approach under their pseudonym of "Isadore Nabi." Their message is that if Watt's systems approach had been used in physics, laws of motion would never have been discovered, and the same is true in ecology. Nevertheless, the somewhat technocratic version of systems ecology has tended to dominate environmental regulators and managers.

Ecological science largely (though not entirely) underwent a shift during the 1930s and 1940s, from a progress-toward-an-ideal (climax community) and harmony of nature view (via the notions of organism and equilibrium) to a disequilibrium view in which many possible end-states are possible for an ecosystem. Despite this direction of development William Morton Wheeler, an expert on social insects at Harvard, emphasized the superorganism concept. During the 1930s, biochemist and physiologist L. J. Henderson, author of the puzzling *The Fitness of the Environment*, led the "Harvard Pareto Circle" (Heyl, 1968). Henderson supported and interpreted the ideas of the early twentieth-century sociologist and economist Vilfredo Pareto using ideas from chemistry and physiology. Pareto was one of the "Machiavellians" in the theory of social elites and was praised by the fascist Mussolini. The circle included the soon-to-be major sociologists Talcott Parsons and George C. Homans, founding sociologist of science Robert K. Merton, and the historian Crane Brinton (whose model of revolution in *Anatomy of Revolution* was influenced by Henderson's version of Pareto and in turn influenced Thomas Kuhn's *Structure of Scientific Revolutions*) among others (see chapter 1). The group encouraged a functionalistic view of society as a kind of self-regulating organism that reaches equilibrium.

During the 1920s Alfred North Whitehead, British mathematician and logician turned Harvard metaphysician, developed a "philosophy of organism," tied to relativity theory and early subatomic physics, that held strong appeal to organismic biologists. Whitehead emphasized the priority of process over enduring substance and the priority of relations over simple qualities (see box 12.2). Significantly, Whitehead also gave a defense of romanticism against the mechanistic worldview of the eighteenth-century Enlightenment in one of the most accessible and influential chapters of his *Science and the Modern World* (1925). At the University of Chicago the ecologists Warder Allee and Alfred Emerson, Thomas and Orlando Park, and Karl Schmidt, who all co-authored a major text in ecology, followed Whitehead's philosophy, as did fellow biologists, the embryologist Ralph Lillie, and Ralph Gerard. Sewall Wright, the Chicago co-founder of mathematical population genetics, though not a member of the Allee group, was also a panpsychist process philosopher, whose views were praised by Whitehead's leading disciple, but who did not much trumpet his views, as they seem to have had little discernable influence on his actual research (Provine, 1986, pp. 95–6). This organismic influence continued through the 1940s but most ecologists were turning to the energetics/economics conception.

Around 1970 (the year of the first Earth Day in the USA) the green movement or ecology movement began as a mass *political movement*, as opposed to a trend within the scientific community. "Ecology" spread from being a division of biology and the preserve of a relatively small number of biologists and conservationists to a mass social movement. The ecology movement largely retained a conception of pristine nature in harmony in a static equilibrium as its ideal. New Age holists such as the ex-physicist Fritjof Capra (1982) appealed to ecology and the green movement to reinforce their holism. On the other hand, the fact that even the population biologist Paul Ehrlich should title his excellent popular summary of scientific population biology *The Machinery of Nature* (1986) shows how far most modern, scientific ecologists are from the organismic or holistic opposition to mechanism as a philosophy. The more radical political ecologists would and do (when aware of it) look askance at the mechanistic viewpoint of many systems and population ecologists as well as at the domination of nature or managerial ethic held in much of systems ecology.

Deep Ecology

Deep ecology is a movement whose name and principles were first formulated by the Norwegian philosopher Arne Naess (1973). Deep ecology claims that the usual approaches in scientific ecology and the environmentalist movement are "shallow," in that they treat nature as an object of human use for human benefit. Deep ecology claims that we have to go further and treat nature as having value in itself, apart from any human use.

The deep ecology movement emphasizes the intrinsic value of wild nature. Deep ecology rejects the viewing of nature as instrumental to human well-being. It rejects the anthropocentric (human-centered) approach to nature. It contrasts as strongly as possibly with the goal of human control of nature found in much Western thought of the past two centuries. Naess's deep ecology has drawn inspiration from the early modern philosopher **Baruch Spinoza** (1632–77), with his completely naturalistic approach to philosophy and his goal of identification of the self not with the selfish ego but with the broadest environment, ultimately the universe. For Spinoza there is only one real substance or thing, the universe as a whole, which he identifies with God (see box 11.2). Some later philosophers of deep ecology use the philosophy of Martin Heidegger to support their position. They draw on Heidegger's turning away from a human-centered or subjective approach to the nature of knowledge and being. They also appreciate Heidegger's conception of the earth as not fully graspable in knowledge and the contrast of earth (or nature) with scientific abstraction (see box 5.1).

For deep ecologists and to a lesser degree many other radical political ecologists, the approach of mainstream scientific and government agency

Box 11.2

Spinoza, Einstein, monism, and holism

The seventeenth-century philosopher Spinoza opposed the mathematician-philosopher René Descartes's sharp division between mind and matter (so-called Cartesian *dualism*). Descartes claimed that there are two fundamental kinds of substance, material (or body, interpreted by Descartes as spatial extension) and mental (or thought). For Spinoza, in contrast, mind and matter, or thought and extension, are two aspects of a single underlying substance whose full nature we do not know, since it is infinite. Spinoza emphasized the bodily parallel to all thought. He also made a deep and detailed analysis of the emotions, which are both bodily and mental. Spinoza has been portrayed as a forerunner of psychoanalytic theories of bodily expression of mental states and of psychosomatic illness, such as those of Freud and of other twentieth-century thinkers who emphasized the physiological and emotional bases of all thought. Spinoza was a thoroughgoing **naturalist**; that is, he denied that there is any aspect of reality that is not part of the natural world. Spinoza was a **pantheist** who identified God with nature ("God or nature"). He was also a **monist** who claimed not only that there is only one *kind* of substance, but also that there is *numerically* only one substance: God = universe.

Albert Einstein was a great admirer of Spinoza. Einstein admired Spinoza's naturalism and sense of awe for and worship of the universe. Einstein, like Spinoza, did not believe in a personal God but had a religious awe for the mysteries of the cosmos. In some of Einstein's more speculative interpretations of General Relativity there is only one thing: space-time. "Things" in the everyday sense (particles) are singularities or warps in space-time. This theory (later called geometrodynamics by John Wheeler) has a strong resemblance to Spinoza's one substance view.

The monism of Spinoza and of geometrodynamics is an extreme form of holism. It claims not only that the whole is prior to the parts but that the parts do not have real existence at the most fundamental level; only the whole system does. Naess and some other ecologists support this form of holism and even relate it to geometrodynamics, as well as to Spinoza and panpsychism (Mathews, 1991, 2003) (on panpsychism, see box 12.2). The whole has properties that the parts do not possess. In many forms of holism, the properties of the whole cannot be fully explained simply on the basis of knowledge of the properties of the parts.

Nevertheless, most biological holists are not monists. That is, the parts do have independent existence even if they are intimately related to one another. The genuine monist, such as Spinoza, denies that the parts have independent reality at all; they are simply "modes" or local modifications of the one real entity or substance (see Dusek, 1999, chapter 1).

ecology is too close to the very technological domination of nature that deep ecologists oppose. The terminology of "managing ecosystems" that many mainstream environmental scientists and agencies use implies this sort of control. Deep ecologists claim that such system management itself is part of the disease, not the cure. Despite this, Naess himself sometimes falls into technocratic language in his presentation of deep ecology.

Ecofeminism

Ecofeminism is a movement combining ecological and feminist concerns. Ecofeminists claim that patriarchy, or male dominance of society, is linked to exploitative and destructive approaches to the environment. In chapter 9 we discussed the gendered metaphors traditionally used to discuss the "man–nature" relationship. From the time of Francis Bacon and the Royal Society until our time, nature has often been portrayed as female, while the investigator or exploiter of nature has been portrayed as male. Unexploited land or timber is called "virgin," while unproductive land is called "barren." Many ecofeminists claim that women by nature have more affinity for and sympathy with wild nature. Others claim that the structure of society – with men in power – structures gender roles and the personalities of men and women such that men are likely to take an attitude of domination over nature, while women tend to have more respect for and are oriented toward the protection of nature.

Karen Warren (2001) enumerates a number of ways in which, she claims, there is a link between patriarchy and anti-ecological attitudes. Some of these are the linguistic connections mentioned above. Others include conceptual connections of hierarchical thinking in terms of the superiority of certain terms traditionally associated with male rather than female attributes. Examples of these are reason/emotion, mind/body, culture/nature, and man/nature, and there are many others. The first is associated with men, and is assumed to be superior, and the latter is associated with women, and claimed to be inferior.

Traditionally, Judeo-Christian religion has emphasized the superiority of humans to nature as well as the priority, both in creation and in rank, within the family of men to women. Standard examples of this are the creation of Eve from Adam's rib in the first version of the Genesis creation story, and St Paul's admonition that women should submit to their husbands.

Ecofeminists, particularly those concerned with the developing nations, note the ignoring of and underemphasis on the contribution of women's work, particularly in agriculture, in the developing world. Schemes for Western aid and development have generally ignored the important role of women's subsistence agriculture, and encouraged industrial and agribusiness development that has traditionally involved male workers.

Ecofeminists have criticized deep ecology, claiming that despite its claims to reject the orientation toward domination of nature and construal of nature solely as human instrument, deep ecology, because of its masculine origins, partakes of the hierarchical and abstract thinking that it claims to reject. Ariel Salleh (1984) notes that founder Arne Naess himself formulates his presentation in terms of the analytic and positivistic philosophy with which he began his career as a philosopher (see chapter 1). Naess speaks of axioms and deductive consequences and of intuitions that need to be made precise. He also compares his approach to general systems theory. Salleh and other ecofeminists see this as a masculinist betrayal of the real implications of deep ecology, which they say would lead away from a formalistic and technocratic approach to knowledge.

Despite these differences, ecofeminists and deep ecologists have more in common with each other than either does with the technocratic and managerial utilitarian approach to environmental and wildlife management. The irony, pointed out by Worster (1977) and Bowler (1992) among others, is that scientific ecology itself, as it becomes more rigorous and quantitative, is enshrined as a professionalized academic subject, and reaches for more influence on government and corporate policy, is generally moving far away from its original romantic and organismic inspirations in a technocratic direction, often in conflict with the personal opinions concerning preservation of wild nature of the ecological scientists themselves. It remains to be seen whether the collaboration of scientists with (or at least having support for) the aims of the green movement will shift the science itself back onto a more holistic path (Bowler, 1992, p. 550–3).

Overpopulation and Neo-Malthusianism

We noted in passing above that some of the holist and monist movements of the beginning of the twentieth century became associated with Nazism. Indeed, Nazism, despite its horrendous murder of millions of members of supposedly "inferior" races based on pseudo-biological theories, also made a major effort to preserve wild areas and species, as part of its "back to nature" and "blood and soil" emphases. Thus, as Bowler (1992, pp. 437, 551) and others have noted, radical ecology has associations with the Nazi right as well as the communal anarchist left.

Another area of political ambivalence of ecological policy is in the area of population limitation. Campaigns against world overpopulation are called neo-Malthusian, after Parson Thomas Malthus's *Essay on Population* (1803). Malthus famously argued that human population grows geometrically, as in the sequence 1, 2, 4, 8, 16, while agricultural production grows arithmetically, as in the sequence 1, 2, 3, 4, 5, such that population vastly outruns food supply. Malthus opposed birth control and thought that the "inferior" lower classes were incapable of exercising the sexual self-restraint that the upper classes did, and hence concluded that the poor will always be with us. Marx and Engels fulminated against Malthus, as they saw him blaming the poor for their poverty rather than the capitalist economic system that kept their jobs scarce and their wages low. Malthus, according to Marx, was "a shameless psychophant of the ruling classes" (Marx and Engels, 1954).

In the twentieth century Marxists such as Paul Baran (1957) criticized ecological neo-Malthusians such as bird biologist William Vogt (1948) by pointing out that the population of Belgium or England was three times denser in the mid-twentieth century than that of "overpopulated" India and twenty times as dense as Sumatra, Colombia, Iran, or Bolivia (Baran, 1957, p. 239). Baran and some other, non-Marxist, economists argued that "overpopulation" was an artifact of bad organization in terms of the landlord system of agriculture in less developed nations (Baran and others saw these nations not as euphemistically "developing nations" but as underdeveloping nations, kept poor by the neo-imperialism of the industrial nations). Mao Zedong, the mid-twentieth-century Communist dictator of China, claimed that "Every mouth comes with two hands attached" (Hertsgaard, 1997) and that rational organization of agriculture could counteract scarcity. (In contrast, since Mao's death and the introduction of capitalist markets in China, draconic forced population limitation measures have been instituted, leading to widespread abortion of female offspring.)

Roman Catholic writers have likewise objected to Malthusianism in order to defend the consequences of the Catholic Church's opposition to birth control and abortion. Although the strictest opposition to contraception (obviously not followed by most Catholics in the industrialized west) is associated with the most conservative aspects of Catholic theology, the Catholic opposition to neo-Malthusianism has often gone along with left-wing "liberation theology" sympathy for the demands of the poor in the developing world.

The scientific fallacies of the early eugenics program were exposed by modern genetics. It turned out that most deleterious genes are recessive, and are carried in a single copy by many people who do not show the disease that the gene causes when present in two copies. Thus negative eugenics (at least before the arrival of genetic screening in the late twentieth century) would not succeed simply by preventing those with genetic diseases from breeding. Furthermore, many illnesses are caused by numerous genes working together, and eliminating "bad" genes turns out to be more difficult than the early eugenicists thought. Once the racial extermination horrors to which Hitler and the Nazis took the program were exposed in 1945, eugenics was almost universally rejected.

After the decline of simplistic early eugenics, many early advocates of limitation of reproduction by the "unfit" and the "inferior" races through eugenics shifted to the apparently more neutral "population limitation" programs sponsored by the Rockefeller Foundation. For instance, Raymond Pearl made the transition in his own lifetime (Allen, 1991). Later neo-Malthusians have still sometimes associated themselves with anti-immigration groups that show racist orientation in their propaganda. Marxist and Catholic opponents of Malthusianism can point out the class and racial biases of many Malthusians, but the issue of the validity of the claim of limits to unchecked population growth remains.

The Stanford University ecological scientist and expert on butterflies Paul Ehrlich has long campaigned for population limitation, writing books with titles such as *The Population Bomb* (1968) and, co-authored with Anne Ehrlich, *The Population Explosion* (1991), and the textbook *Population, Resources, Environment* (1972). Ehrlich was also the moving spirit of the organization ZPG, or Zero Population Growth. Ehrlich himself is explicitly anti-racist, having authored with S. Shirley Feldman *The Race Bomb* (1978), and considers himself a social democrat.

However, some have criticized Ehrlich's early blaming of poverty and pollution solely upon population growth in *The Population Bomb* and ZPG slogans, and claimed that he was blind to the economics of capitalism as well as implicitly racist with respect to the high-population, less developed world. Since his earliest book on the topic Ehrlich has greatly qualified and elaborated on his position with respect to issues of equity and the role of population *vis-à-vis* industrialization, underdevelopment, and other factors, while emphasizing the untenability of unchecked (or, indeed, any further) world population growth.

Nevertheless, biologist Garrett Hardin, whose politics are explicitly on the right, advocates more ruthless Malthusianism of the older sort. (His spouse stabbed one of the Science for the People members with her knitting needle at an American Association for the Advancement of Science meeting (Kevles 1977).) In his "lifeboat ethics" Hardin (1972, 1980) has counseled against food aid in famines in Africa, claiming that this simply maintains the high population that will cause more famines.

The biologist and ecological activist Barry Commoner (1975) summed up the criticism of these views in an article entitled "How poverty breeds over-population (and not the other way around)." Commoner and some other economists and demographers claim that pre-industrial, agrarian societies encourage large families and that economic development leads to progress-ively lower birth rates. In the case of pollution, it has often been pointed out that the developed industrial countries produce many times their share of solid waste and greenhouse gases, though developing nations such as China will soon dwarf others as they industrialize. The USA uses 25 percent of the world's resources with only 4 percent of the population, while the industrialized countries (the USA, Europe, and Japan) use 80 percent of world resources. China in 2004 had one million automobiles. If privatization of transportation leads to widespread gasoline-powered automobile use com-parable to that in the West, China will have many hundreds of millions of automobiles and enormous pollution. It seems unlikely, to say the least, that China and India can develop in the wasteful and energy-intensive manner of the presently industrialized countries, although there are utopians who claim that technological advances will allow this to happen.

With respect to famines in Africa, some demographers and sociologists on the left have claimed against Hardin and others that civil wars and the breakdown of transportation of food are often the cause of famines in Ethi-opia and Sudan, not the absolute inability of the land to support the popula-tion (Downs et al., 1991; Reyna and Downs, 1999). In other parts of Africa, the monoculture (single crop agriculture) practiced by multinational corporations, which replaces local subsistence farming, and the exporting of food to more affluent European markets even during famines have been claimed to exacerbate African food shortages (Lappé et al., 1979; Lappé and Collins, 1982).

Ecologists have long argued that world population needs to be limited. It would seem indisputable that there is some limit on human population in terms of the carrying capacity of the planet. Nonetheless, some technological optimists as well as Marxists and Roman Catholic theorists (for the very different reason of opposition to artificial birth control and abortion) have disagreed even with that claim. In the 1960s, architect, inventor, and utopian Buckminster Fuller once claimed that more than the whole world population could fit onto Manhattan Island "with enough room to dance the twist" (although a telephone-booth sized living space would seem to be somewhat constraining). Fuller, who thought his geodesic dome would end poverty, believed that a less wasteful technology would allow vast population increase. He believed his design science thinking had refuted Malthusianism.

Between the extremes of Mao, Bucky Fuller and some Roman Catholic writers who deny there is any problem of overpopulation, and the simplistic but easy to communicate early Ehrlich message that "population = poverty plus pollution," a number of writers have attempted to combine a demand for world population limitation with demands for equity between nations and regions and an avoidance of blaming all problems of poverty and pollution on population alone.

However, the fact that overpopulation theory or neo-Malthusianism has found support among both left-wing, communal anarchists and right-wing racists and class elitists shows the complexity of any simple association of ecological demands with a single political position.

Sustainability

The metaphor and language of **sustainability** has become the central manner of expressing ecological concerns about the economy and technology today. The very ambiguity of the term has allowed many varied groups and individuals with diverse theories and programs for health or ecological survival to find common ground. Sustainability sounds less radical than such movements as anarcho-communism and bioregionalism, although its implications, if taken seriously, may be equally or even more radical. At present, sustainability is sufficiently broad in concept to accommodate ecological radicals, government regional planners, neo-liberals, and corporations offering sustainable products. There is even a Dow Sustainability Index, tracking the stocks of corporations that claim to contribute to sustainability.

Attempting to formulate a definition of sustainability leads to many of the issues and problems that we had when attempting to give a definition of technology (see chapter 2). One website lists some 27 definitions of sustainability. A regional planning commission declares "The definition of sustainability depends on who [sic] you talk to." This statement seems to accept the conventional notion of definitions as arbitrary, or the extreme idiolect version of descriptive definitions, where every person has his or her own personal definition. There also are a number of what might be called "feelgood" definitions of sustainability that say sustainability is the simultaneous maintenance of biodiversity, economic development, and personal growth for all citizens. This sounds great, if it can be achieved.

Despite the variety of definitions available, there are a few core factors common to many definitions, or at least present in many definitions. One is clearly the maintenance of resources and environmental integrity for future generations. One of the earliest and most influential definitions of sustainability is found in what is informally called the "Brundtland Report" and officially titled *Our Common Future*. The World Commission on Environment and Development defined sustainability as the project to "meet present needs without compromising the ability of future generations to meet their needs" (WCED, 1987). The theses of this report have since been elaborated and clarified in subsequent conferences, such as the Rio conferences on biodiversity, starting with the United Nations Conference on Environment and Development at Rio de Janeiro in June 1992.

Some definitions of sustainability make this the sole characteristic of sustainability, the "seven generations" test attributed to the Seven Nations, known to the white settlers as the Iroquois tribe of Native Americans. The notion of "sustainable agriculture," which pre-dates and is a source of the general notion of sustainability, emphasizes that the resources of the land, the fertility of the soil, should not be decreased by its agricultural use. The leaching of soil by irrigation, as in the salting up of the canals in ancient Mesopotamia (Iraq), the loss of soil, as in the amazing amount of topsoil that is lost in the North American Midwest by flowing into the Mississippi River, and the loss of nutrients of the soil through intensive farming are all disturbing examples of what sustainable agriculture wishes to avoid. Some of the agricultural and biological definitions of sustainability are simple input/

output definitions, rather like the physical definitions of mechanical efficiency (see chapter 12).

These simple input/output bio-efficiency definitions are rejected by many advocates of sustainability, because they ignore two other factors that these critics of such definitions deem essential to sustainability. First, there is the maintenance of biodiversity. This is not simply the maintenance of agricultural productivity, but the maintenance of the biodiversity of the surrounding ecosystem. Second, the input/output form of definition, even with concern for future generations and support for maintaining biodiversity, ignores the human factor. The comfort, health, and well-being of the human population must be sustained. Furthermore, the human potential for development and fulfillment should be maintained.

"Sustainable development" is another sustainability term that was devised prior to the notion of "sustainability" in general. There has been a series of United Nations Commissions on Sustainable Development since 1992.

This leads us to an attempt to formulate a comprehensive definition of sustainability. Sustainability includes the: (a) maintenance of resources, particularly the use of renewable resources; (b) passing on of resources, the environment, and social benefits for future generations; (c) preservation of biodiversity and the integrity of the environment; (d) maintenance of technological and economic development, enhancing the well-being of the human population; and (e) fostering and enhancing of a comfortable and fulfilling lifestyle for the human inhabitants.

Sustainability combines the advocacy of technological and economic development with biodiversity and the use of renewable resources. What is not clear is that all these desirable things can be simultaneously maintained. Supporters of sustainability are optimists, in that they think that all these desirata can be achieved.

Study questions

1   Is the term "Luddite" historically accurate when used to describe contemporary critics of technology?
2   Do you think that holism is a valuable approach to nature? Is holism a valuable approach to science or is the analytical and atomistic approach the only viable one?
3   Is deep ecology a broader and more decent ethic than human-centered views, or is it inhumane and lacking proper consideration for human beings?

4   Do you think there are conflicts between the results of ecological science and the goals of the ecology movement? If so, why? If not, why not?
5   Do you think that overpopulation is the major source of problems of food shortages, pollution, and poverty?
6   What is, in your opinion, the best definition of sustainability? Why is it superior to others?

12

Social Constructionism and Actor-network Theory

Social constructivism has become a common approach to a variety of topics in a variety of fields of the humanities and social sciences. Its heyday has been since the 1980s, even though the basics of constructivist philosophy, mathematics, and psychology have deep historical roots. The theme of constructivism goes back at least several centuries in philosophy and has been tied to fields as diverse as the theory of knowledge, mathematics, developmental psychology, and the theory of history and society (see box 12.1).

Sociology of Knowledge as a Prelude to Constructivism

Around the beginning of the twentieth century sociologists influenced by Kant, such as Georg Simmel, developed notions of society that emphasized the construction of the social world. Georg Lukács (1885–1971), a Hungarian, studied under Max Weber the sociologist (see chapter 4), Wilhelm Dilthey the hermeneuticist (see chapter 5), and Georg Simmel (1858–1918), a sociologist associated with the historically oriented neo-Kantians, who wrote in *Philosophy of Money* (1900) on, among many other social topics, the pervasive influence of money on the rise of subjectivity in the modern era. Lukács, once he moved, literally almost overnight, from existentialism and neo-Kantianism to communism during the social crisis in Hungary at the end of the First World War, presented a view of society as constructing the categories of the world, with the claim in his *History and Class Consciousness* (1923) that capitalist and communist society would categorize and understand the world differently. In many respects, Lukács rediscovered in his own thought the ideas of the early Marx a decade before they were finally published.

Box 12.1

History of constructivism in philosophy and other fields

Constructivism is a tendency in the philosophy of the past few centuries. Probably the earliest proponents of the claim that our knowledge is constructed were **Thomas Hobbes** (1588–1679) and **Giambattista Vico** (1668–1744). Both of these philosophers claimed that we know best what we make or construct. Hobbes claimed that mathematics and the political state were both constructed by arbitrary decision. In mathematics and science, the arbitrary decision is stipulative definition (see chapter 2). In society, the decision was the subordination of oneself to the ruler in the social contract. Vico claimed that we know mathematics and history best because we construct both. For Vico history is made by humans in collective action.

The major source of the varied ideas of constructivism in many fields is **Immanuel Kant** (1724–1804). Kant (1781) held that mathematics is constructed. We construct arithmetic by counting and construct geometry by drawing imaginary lines in space. Kant also claimed that we construct concepts in mathematics, but that philosophical (metaphysical) concepts are not constructed, but dogmatically postulated in definitions. For Kant, the source of the constructive activity is the mind. Various faculties or capacities are part of the makeup of the mind. Kant differed from the British empiricists in that he emphasized the extent to which the mind is active in the formation of knowledge. Although sensation is passive, conceptualization is active. We organize and structure our knowledge. Through categories we unify our knowledge. Kant compared his own innovation in theory of knowledge to the "Copernican Revolution" (the astronomical revolution of Copernicus that replaced the earth as the center of the solar system with the sun as center). Some have suggested that, given that Kant makes the active self the center of knowledge, his revolution is more like the Ptolemaic, earth-centered, theory of astronomy.

After Kant a variety of tendencies in philosophy furthered constructivist notions. While Kant claimed that we structure our knowledge or experience, he held that there is an independent reality, **things in themselves**, which we cannot know or describe, because all that we know or describe is structured in terms of the forms of our perceptual intuition and the categories of our mind. We can know that things in themselves exist, as the source of the resistance of objects to our desires, and the source of the input to our passive reception of perceptual data. But we cannot know anything about the qualities or characteristics of things in themselves. Gottlieb Fichte (1794), much to Kant's horror, proposed that since the thing in itself is inaccessible, and we cannot say anything about it, philosophy should drop the thing in itself and hold that the mind posits or creates not simply experience but reality as such. For Fichte the mind "posits" reality, and its positing is prior even to the laws of logic. Within later constructivism there are differences between those who claim merely that our knowledge is constructed, as did Kant, and those who claim, like Fichte, that objects and external reality themselves are constructed.

Hegel (1770–1831) added a historical dimension to Kant's categories. Kant had presented the categories as universal for reason as such and basically unchanging. Kant claimed that even extraterrestrials and angels would share our categories. Hegel added an emphasis on historical development to philosophy. Hegel claimed that the categories develop through time and history. Even in logic, there is a dialectical sequence of categories. For instance, Hegel began his *Logic* (1812–16) by generating non-Being from Being and then producing the synthesis of Becoming. In his *Phenomenology of Mind* (1807) Hegel developed the categories of knowledge and ethics in a quasi-historical manner. Ancient Greece and Rome, the French Revolution, and more recent Romanticism furnish examples of the sequence of ethical and social categories. Most later Hegelians in Germany and Italy treated Hegelianism as a fully historical philosophy.

Marx and Engels (1846) claimed that the construction of categories was not a purely mental sequence, but was a material process of actual social and historical activity of production in the economy. For Marx, real, social history, not an idealized history of spirit, generates the frameworks (ideologies) in terms of which people understand the world.

The neo-Kantian school revived Kant's ideas in the late nineteenth century. One branch of the neo-Kantians (the Marburg School) emphasized the construction of lawful (nomothetic) scientific and mathematical knowledge. Another branch (the Southwest German School) focused on the construction of historical and cultural knowledge of unique individuals (idiographic knowledge) in the humanities. The nomothetic versus idiographic dichotomy is manifest in the later opposition of the logical positivists and the hermeneuticists (see chapter 1 on positivism and chapter 5 on hermeneutics).

In the late nineteenth and early twentieth centuries, constructive ideas of mathematics were developed into detailed philosophies of mathematics and programs for the rigorous building up of systems of mathematics. Henri Poincaré in France (who also was granddaddy of parts of chaos theory – see box 6.2) and Jan E. Brouwer in Holland claimed that mathematics is built up from the ability to count (Brouwer) and the principle of mathematical induction (Poincaré), and that these notions went beyond purely formal logic (Poincaré, 1902, 1913; Brouwer, 1907–55). **Constructive mathematics** is a growing minority current within mathematics (Bridges, 1979; Rosenblatt, 1984), and has gained some renewal from theories of computers and computation (Grandy, 1977, chapter 8). Several of the leading theorists in the constructivist approach in education began in mathematics education (von Glasersfeld, 1995; Ernest, 1998).

In the early twentieth century, the British logician Bertrand Russell (1914) and the logical positivist Rudolf Carnap (1928) developed the notion of the **logical construction** of the world. According to Russell's logical construction doctrine, all of our factual knowledge is based on data from direct sense experience (or sense data). Physical objects are simply patterns and sequences of these sense data. What we commonsensically call a physical object is, in terms of rigorous knowledge, a logical construction from sense data. The various perspectives we have or experience are systematically combined and organized to give us the notion of the physical object. This form of constructivism is different from Kant's constructivism and is a logically developed version of British empiricism. It is interesting, however, to note that constructivist themes were very much a part of the logical positivist tradition that social constructivists in contemporary philosophy of technology think they are rejecting. At least one contemporary survey and evaluation of social constructivism includes logical constructivism in its scope (Hacking, 1999). Ironically, Hume's purely logical criticism of induction and causality led to logical positivism, but his appeal to "habit and custom" in explaining the psychological reasons why we believe in induction and causality strongly resemble social constructivist accounts.

Kantian constructivist ideas also found their way into psychology in **Jean Piaget's** (1896–1980) theories of the growth of knowledge in the child. Piaget has very strongly Kantian approaches to knowledge, especially in his early works (Piaget, 1930, 1952). Unlike Kant, Piaget believes that the categorization or organization of knowledge develops and changes with the growth of the individual, just as Hegel thought it developed historically in society. For Piaget, the categories of thing and of conservation can be shown to develop through various stages. The Russian psychologist Lev Vygotsky (1896–1934) developed a theory of cognitive development that emphasized more than Piaget the social dimension of the development of the child's conceptual framework (Vygotsky, 1925–34b).

Because of Lukács's training in Kant and Hegel he was able to rethink the conceptual moves that Marx himself had made from earlier German philosophy. However, Lukács's conclusions were more spiritual and idealistic than those of Marx. This earlier Lukács, who later claimed self-critically in 1967 that he had earlier "out-Hegeled Hegel," claimed that the workers could achieve absolute knowledge after the communist revolution. This is because the workers are positioned to grasp the essence of capitalist production and the "reification" it produces. Marxism is the theory of the movement of the workers. Marxism is true because it reflects the absolute knowledge that the workers will achieve. However, the reason we can be sure that the workers' movement will achieve communism is that Marxism says so, and is true. The argument is circular. The victory of the workers leads to the viewpoint that vindicates Marxism's truth, but the truth of the Marxist theory of history guarantees the victory of the workers. Lukács himself later gave up this circular justification, but merely allied himself with dogmatic "scientific" Marxism-Leninism. Without this later dogmatism, one is left with relativism again.

From Marxist concepts of ideology (minus the partisanship of Marxist politics) the **sociology of knowledge** developed. Sociology of knowledge, such as that of Karl Mannheim (1893–1947), a non-Marxist sociologist very influenced by Lukács, claims that knowledge is conditioned and constituted by the social position and role of the knower, but does not assign any special role to the working class in grasping truth (see the discussion of Mannheim in chapter 1). Mannheim (1929) claimed that his "relationism" overcame relativism. Relationism, in various formulations, adjudicates between the different ideological standpoints by relating them to one another and synthesizing them. It also admits its own socially conditioned nature, supposedly thereby escaping the naiveté of ideological positions. Mannheim also sometimes appealed to the "free floating intelligentsia" as being less biased than other classes, such as capitalists and workers, and claimed that the viewpoint of intellectuals was more accurate than that of other groups. Mannheim's solutions are generally conceded to have failed to fully overcome relativism. Nevertheless, his analyses of various movements of thought in relation to their social context and conditioning launched the sociology of knowledge.

A work that greatly influenced later social constructivism in the English-speaking world is **Peter Berger** and **Thomas Luckmann**'s *The Social Construction of Reality* (1966). This work, subtitled *A Treatise on the Sociology of Knowledge*, presented ideas akin to those of Lukács and Mannheim in a politically neutral and easily accessible English language formulation. Berger and Luckman concentrated on the constitution of knowledge of society, but their ideas could be adapted to knowledge of the natural world by sociologists of scientific knowledge (see the discussion of sociology of scientific knowledge in chapter 1).

What Is Constructed? The Varieties of Constructivism

Social constructionism (or constructivism) is perhaps the dominant tendency in the sociology of scientific knowledge (SSK). Social constructionism was applied later to technology and is becoming a major tendency in the social theory of technology. Social construction of technology (SCOT) is somewhat less controversial than the social constructivist theory of science. This is because SCOT does not have to confront questions from the theory of knowledge in their most controversial form, which SSK does have to confront. The claims that artifacts or devices are constructed and that theories are constructed are not metaphysically controversial. The addition of the adjective "social" to the account of the construction makes it descriptively differ from accounts of individual construction of artifacts and theories, but does not raise fundamental issues about reality (though the latter raises issues about the nature of knowledge). SSK has to face the apparently more philosophically speculative claim that scientific facts and objects are socially constructed.

The most philosophically controversial aspect of social constructionist SSK is the claim that facts or physical objects and events are social constructed (see chapter 1). This claim can be made to sound innocuous by saying that scientific "facts" are simply those statements accepted by the scientific community and that scientific "objects" are simply those concepts or names of objects that are accepted in scientific theories. However, if *more* is claimed than that our *belief* in facts or the existence of objects is constructed

– say, that there is *no* difference between our belief in facts or objects and the facts or objects themselves – then one is led into philosophical questions about the degree of objectivity or subjectivity of our knowledge of the world. The question arises as to whether objects, events, and facts exist independently of knowledge or construction of them. Sociologists of scientific knowledge sometimes claim to be merely doing sociology and not making philosophical claims, but this flies in the face of the obviously philosophical claims that partisans of SSK sometimes make.

SCOT seems more reasonable and less controversial than the social constructivism of nature or physical reality, in that technological artifacts are indeed made. Devices, artifacts, or inventions are literally physically constructed. Insofar as their construction involves the collaboration of people or even the utilization of techniques, viewpoints, and facts borrowed from others, the construction is social in an obvious sense. Furthermore, theories and models are conceptually constructed. Insofar as concepts of others are utilized, past or contemporary, this construction is also social. Furthermore, because of the active, manipulative nature of technology, techniques, guiding principles, concepts, and theories are embedded in the physical construction of technological artifacts. The interpenetration of socially constructed concepts and socially constructed devices is clearly evident in technology. This is more obvious than in science, given the tradition of understanding the latter as purely theoretical knowledge.

Similarly, the claim that social groups and social institutions are socially constructed is a much less controversial aspect of social constructionism than the claim that the objects and facts of science are socially constructed. For instance, the philosopher John Searle, in *The Construction of Social Reality*, argues strenuously against the social construction of physical objects or the objects of science, but develops an account of the social construction of institutions (Searle, 1995). Searle defends a notion of "brute facts" concerning the physical world in strongest contrast to the notion of socially constructed physical facts. However, he develops at length the notion of social facts as socially constructed. (Searle appeals to "speech acts," which are "performative"; that is, create social relations or institutions. Examples are "I do" in marriage or "I promise" in an agreement or contract.)

The technological systems approach to technology emphasizes that technological systems involve both physical artifacts and social relations of producers, maintainers, and consumers of technology. Hence the social production of artifacts is intertwined with the social production of such groups. Thus, in the systems approach, the sharp distinction that Searle and

204

other anti-constructionists with respect to science would make between the purely physical technological brute facts about devices and the purely social and constructed facts about social relations and institutions becomes blurred.

The controversial claims of the social construction of science, the claims that facts, realities, and nature are socially constructed, are much less central to the consideration of technology. However, one aspect of social construction of technology that resembles the social construction of facts is the claim that the effective functioning of technological devices is socially constructed. That is, effective function, or "working," is taken not as a physical given but as a social arrangement by the constructionist approach. In engineering the efficiency of a device is taken to be a purely quantitative ratio of the energy input and energy output. However, in practice, whether a device is considered to "work well" is a product of the character and interests of a user group.

The same device may be considered effective by one group and ineffective by another. Bijker (1995), in his study of the bicycle, shows how the large wheel bicycle was considered functional and useful by youthful athletic users, but dangerous and unstable by other, everyday users.

Social constructionists emphasize that a technological artifact is only the totality of meanings attributed to it by various groups. A pure social constructionism of technology leads to the view that there is no underlying, neutral physical device, only differing meanings and evaluations attributed to it by various groups. This leads back to the metaphysical issues about the reality of objects independent of human attitudes to them. For instance, the realist would claim that the inputs and outputs of energy and work are physical realities, even if different groups might consider the same device "efficient" or "inefficient" depending on their criteria, goals, and needs. However, social constructionists with respect to technology forefront such metaphysical issues far less than do social constructionist accounts of science.

Winner's Criticisms of Social Constructionism

Langdon Winner (1993) notes that social constructionism shares some of the features and weaknesses of pluralism in political science. The pluralists oppose those theories of rule, such as Marxism and power elite theory, that claim there is a ruling class or a power elite that rules society. Pluralists deny that there is a ruling elite and claim that political decisions are the product of the interaction of a number of interest groups. The critics of pluralism note that some of these groups, such as the very wealthy, are far more influential on social policy than others, such as the very poor. Furthermore, the pluralists study conflicts among proposed alternative policies but neglect to note the possible policies that never are on the agenda, because they are excluded by the terms of the debate. According to the critics, the most powerful groups frame the alternatives in the debate and set the agenda.

Winner claims that the social constructionists emphasize the diversity of groups that influence technological development without noting how some of those groups dominate the development of the technology and others have practically no say. For instance, the interests of owners and managers determine the design of factory technology to the almost complete exclusion of the desires and interests of laborers (Noble, 1984).

Winner's criticism of social constructionism above is not in itself a criticism of the general or abstract social constructionist thesis. That is, technological artifacts could still be socially constructed, but constructed entirely or primarily in terms of the goals and values of one dominant group. Nevertheless, in practice those who dub themselves social constructivists do oppose Winner's claims and criticism. Social constructivists emphasize interpretive flexibility and the great variety of meaning attributed to what the naive observer might think to be the "same" technical device.

The response of Steve Woolgar and Mark Elam to Winner's criticism shows the extent to which liberal or even libertarian pluralism is incorporated into much social constructionism (Elam, 1994). Winner uses the example from Robert Caro (1975) of New York City planner Robert Moses's building of bridge overpasses on highways to Long Island beaches high enough for private autos but too low for public buses to travel under them. Caro and Winner claim that Moses did so specifically to prevent poor New Yorkers, particularly African Americans, from using the public beaches. Woolgar and Elam emphasize that a variety of other interpretations are possible. Winner asserts against Woolgar that there is a single, true account of Moses's motives and goals, even if we do not know what it is, hence implicitly appealing to a kind of metaphysical realism. (One could bolster the claim that Moses attempted to exclude African Americans from public facilities such as beaches by the further fact that Moses ordered public swimming pools to be kept at cold temperatures, on his often expressed but mistaken belief that African Americans would not swim in cold water.)

Although Winner castigates Woolgar's approach as apolitical, while Elam defends Woolgar's position as a kind of liberal or libertarian irony, Wiebe Bijker (1995, p. 289) emphasizes the political implications and applications of social constructionism. Bijker notes, as have many constructivists, that recognition of the social nature of closure or consensus with respect to technological devices and their "working" can free critics of a technology from resignation to technological determinism. Nonetheless, social constructionism does not automatically imply a leftist politics. Environmentalists who appeal to "hard scientific facts" to bolster their case may wish to deny the socially constructed nature of science and technology. Similarly, defenders of the status quo could use insights from social constructivism to further their own strategies of power and persuasion. For instance, some conservative social constructionists claim that citizens' objections to air pollution are merely primitive taboos and purification rituals (Douglas and Wildavsky, 1982). Social constructivism can be used for political purposes, but it does not predetermine the political positions in support of which it can be used.

Another criticism of social constructivism that Winner raises is that it emphasizes the creation and acceptance of technology but not the impact of the technology. Social constructivism of science is mainly concerned with the creation of theories and experimental observations. Taking as a model the earlier work on science, social constructivists working on technology use the portrayal of the construction of theory and data as a model for the construction of technology. Social constructivists might reply that the "impact" of the technology is treated in the systems of meaning that groups attribute to the technology. However, the emphasis of the constructivists is on the production of the technology itself, not on the broader social changes in institutions and attitudes brought about by the incorporation of the technology.

Actor-network Theory as an Alternative to Social Construction

Bruno Latour, although often associated with social construction because of the subtitle of his *Laboratory Life*, in its first edition, *The Social Construction of a Scientific Fact*, became a critic of social constructionism. (The second edition of the book removed the word "social" from the subtitle, claiming that the emphasis on the social was misleading.) Latour, Michel Callon, and John Law became proponents of **actor-network theory**. In actor-network theory the participants include human actors as well as laboratory animals and inanimate objects. No special priority is given to human society. Latour refers to the elements as **actants**, including non-human living things and physical objects. Actants are recruited into a network. Social construction, following the earlier Latour, may speak of individual researchers being recruited to support a position, but actor-network theory also speaks of laboratory organisms and apparatus being recruited.

Latour (1992) criticizes social constructionism for overemphasizing the power of the human mind or society in constructing artifacts and nature. Social constructionism is a variation on the Kantian (see box 12.1) notion of the mind organizing the world, even if the abstract, universal mind of Kant is replaced by a group of socially interacting individuals in social constructionism. Actor-network theory does not wish to prejudge the relative power or influence of any of the actants.

Furthermore, according to Callon, the boundaries between the technological system and the rest of society or the rest of the universe do not have to be delimited. The technological system cannot be neatly separated off from the rest of society or nature.

Law emphasizes that the apparently purely physical engineer is really what Law calls a **heterogeneous engineer**, who engineers society as well as physical artifacts. The engineering of artifacts should not be separated as "engineering" from the changing of society that is involved in introducing a new technology, such as the automobile or the electric power network.

Box 12.2

Process philosophy

Actor-network theory has a number of resemblances and explicit affiliations with **process philosophy**. Process philosophy is a collective term used for a number of philosophical systems around the turn of the twentieth century. Process philosophy stems from the works of the Frenchman Henri Bergson (1911), the Englishmen Samuel Alexander (1916–18) and Alfred North Whitehead (1929), and in part the more metaphysical aspects of the American pragmatist philosophers Charles S. Peirce (1839–1914), William James (1842–1910), John Dewey (1859–1952), and George Herbert Mead (1863–1931). Process philosophers were influenced in part by the theory of biological evolution, although some, like Bergson (1859–1941) and Peirce, rejected Darwin's natural selection version of evolution for a more purposive evolution. Einstein's Special Theory of Relativity influenced some process philosophers, such as Mead (1932) and particularly Whitehead (1922).

Process philosophy holds that the ultimate entities in the universe are not enduring things or substances but processes. Its oldest precursor in

Western philosophy is the pre-Socratic philosopher Heraclitus, who famously claimed that one "cannot step in the same river twice," and that "there is a new sun every day." Whitehead's process philosophy (as well as that of James and Mead) also makes relations central. Rather than things or substances, relations are the fundamental stuff of reality (basic metaphysical elements). According to James, against Kant, we perceive relations, rather than combining bits of sense experience with our mind. In this, James resembles, and actually influenced, phenomenology (see Gurwitsch (1964) and chapter 5). Whitehead also holds the doctrine of **panpsychism**, in which the ultimate elements of reality are not material particles but psychic entities, in Whitehead's scheme elements of feeling. Latour, in his later work and talks, has made frequent reference to Whitehead's panpsychism.

Latour and actor-network theorists' designation of the elements of their networks as actants, not distinguishing human actors from physical objects usually considered inorganic and inert, has some resemblance to Whitehead's panpsychism. A difference is that traditional panpsychists such as the seventeenth-century mathematician-philosopher Leibniz (1642–1727) and the twentieth-century Whitehead only attributed mental (Leibniz, 1714) or feeling (Whitehead, 1929) properties to what they considered to be organic units. Aggregates, such as a table or a rock, were not attributed consciousness or feeling, though molecules or organs might be.

Whitehead's emphasis on relations clearly fits with actor-network theory, where the network or, better, the production of the network is central and is prior to or at least not built up from its atomistic individuals. Also, Latour (1988) has drawn (controversial) analogies between Einstein's Special Theory of Relativity and the relational and relativistic approaches in the social sciences.

Interestingly, two other leaders of science/technology studies, Donna Haraway and Andrew Pickering, have also recently made reference to Whitehead and made use of his ideas. Haraway and Pickering share with Latour an emphasis on the hybrid or intermediate beings, as in Haraway's use of human–machine combinations, such that intentionality is not attributed to humans alone (Haraway and Pickering, in Ihde and Selinger, 2003). Pickering, oddly, gets his knowledge of his fellow British thinker Whitehead (who had interests in mathematical physics like Pickering) indirectly via the postmodern French theorists Gilles Deleuze (1966, 1988) and Paul Virilio (2000). It would be better to eliminate the middle men.

Actor-network theory has a number of resemblances and explicit affiliations with **process philosophy**, particularly that of the British-American mathematician-philosopher Alfred North Whitehead, to whom Latour in his more recent work often refers (see box 12.2).

Study questions

1 What versions of social constructionism do you think are correct or incorrect? Does the "social construction of facts and artifacts" correctly locate both as socially constructed?

2 Do you think that social constructionism is more reasonable as applied to science to technology?

3 Do Winner's criticisms of social constructionism hit their mark? For instance, is social constructionism a kind of pluralism that ignores the dominant interests that really direct the development of technology?

4 Do you think a case can be made for the claim that the "efficiency" of technological devices is a social construction and not simply a matter of measuring input and output in terms of physical quantities?

Bibliography

Achterhuis, H. (ed.) (2001) *American Philosophy of Technology: The Empirical Turn* (trans. R. Crease). Bloominton: Indiana University Press.

Ackermann, R. J. (1985) *Data, Instruments and Theory: A Dialectical Approach to the Understanding of Science*. Princeton, NJ: Princeton University Press.

Ackroyd, P. (1996) *Blake*. New York: Norton.

Adas, M. (1989) *Machines as the Measure of Men: Science, Technology, and the Ideologies of Western Dominance*. Ithaca, NY: Cornell University Press.

Adorno, T. W. (1998) *Critical Models: Interventions and Catchwords* (trans. H. W. Pickford). New York: Columbia University Press.

Agassi, J. (1971) *Faraday as a Natural Philosopher*. Chicago: University of Chicago Press.

Agassi, J. (1981) *Science and Society: Studies in the Sociology of Science*. Dordrecht: Reidel.

Agassi, J. (1985) *Technology: Philosophical and Social Aspects*. Dordrecht: Reidel.

Agre, P. (1997) *Computation and Human Experience*. Cambridge: Cambridge University Press.

Agre, P. and Chapman, D. (1991) What are plans for? In P. Maes (ed.), *Designing Autonomous Agents: Theory and Practice from Biology to Engineering and Back*. Cambridge, MA: MIT Press.

Alcoff, L. and Potter, E. (eds) (1993) *Feminist Epistemologies*. London: Routledge.

Alexander, S. (1916–18) *Space, Time and Deity*. New York: Humanities Press (1950).

Allen, G. (1991) Old wine in new bottles: from eugenics to population control in the work of Raymond Pearl. In K. Benson, J. Maienschein and R. Rainger (eds), *The Expansion of American Biology*. New Brunswick, NJ: Rutgers University Press.

Allen, S. G. and Hubbs, J. (1980) Outrunning Atalanta: feminine destiny in alchemical transmutation. *Signs: Journal of Women in Culture and Society*, 6(2), 210–21.

Althusser, L. (1966) *For Marx*. London: New Left Books.

Arditti, R., Klein, R. D. and Mindin, S. (eds) (1984) *Test-tube Woman: What Future for Motherhood?* Boston: Routledge.

Arendt, H. (1929) 1996 *Love and Saint Augustine* (ed. J. V. Scott and J. C. Stark). Chicago: University of Chicago Press (1996).

BIBLIOGRAPHY

Arendt, H. (1958) *The Human Condition*. Chicago: University of Chicago Press (2nd edn, 1998). Selection in Scharff and Dusek, pp. 352–68.

Arendt, H. (1964) *Eichmann in Jerusalem: A Study in the Banality of Evil*. New York: The Viking Press.

Aristotle (1985) *Nichomachean Ethics* (trans. T. Irwin). Indianapolis: Hackett (selection in Scharff and Dusek, pp. 19–22).

Aron, R. (1962) *The Opium of the Intellectuals*. New York: Norton.

Ashman, K. M. and Baringer, P. S. (eds) (2001) *After The Science Wars*. London: Routledge.

Atwell, W. S. (1986) Some observations on the seventeenth century crisis in China and Japan. *Journal of Asian Studies*, 14(2), 223–44.

Ayer, A. J. (ed.) (1959) *Logical Positivism*. Glencoe, IL: Free Press.

Bacon, F. (1620) *Novum Organum* (trans. and ed. P. Urbach and J. Gibson). Chicago: Open Court Publishing (1994) (selection in Scharff and Dusek, pp. 29–31).

Bacon, F. (1624) *The New Atlantis*. In *The New Atlantis and The Great Instauration* (trans. J. Weinberg). Wheeling, IL: Harlan Davidson (1989). (selection in Scharff and Dusek, pp. 31–4).

Bagdigian, B. H. (2004) *The New Media Monopoly*. Boston: Beacon Press.

Bailes, K. E. (1978) *Technology and Society under Lenin and Stalin: Origins of the Soviet Technical Intelligentsia, 1917–1941*. Princeton, NJ: Princeton University Press.

Baran, P. (1957) *The Political Economy of Growth*. New York: Monthly Review Press.

Barlow, J. P. (1996) A declaration of independence of cyberspace. Electronic Frontier Foundation (www.eff.org/~barlow).

Baudrillard, J. (1995) *The Gulf War Did Not Take Place* (trans. P. Patton). Bloomington: Indiana University Press.

Bell, D. (1960) *The End of Ideology: On the Exhaustion of Political Ideas in the Fifties*. Glencoe, IL: Free Press.

Bell, D. (1973) *The Coming of Post-industrial Society: A Venture in Social Forecasting*. New York: Basic Books.

Benbow, C. P. and Stanley, J. C. (1980) Sex differences in mathematical reasoning ability: Fact or artifact? *Science*, 210(4475), 1262–4.

Berger, P. and Luckmann, T. (1966) *The Social Construction of Reality: A Treatise on the Sociology of Knowledge*. Garden City, NY: Doubleday Anchor.

Bergmann, G. (1958) *Philosophy of Science*. Milwaukee: University of Wisconsin Press.

Bergmann, P. G. (1974) Cosmology as a science. In R. J. Seeger and R. S. Cohen (eds), *Philosophical Foundations of Science*. Dordrecht: Reidel.

Bergson, H. (1911) *Creative Evolution* (trans. A. Mitchell). New York: H. Holt and Co.

Berle, A. and Means, G. (1933) *The Modern Corporation and Private Property*. New York: Macmillan.

Beveridge, W. I. B. (1957) *The Art of Scientific Investigation*. New York: Norton.

Bijker, W. E. (1995) *Bicycles, Bakelites and Bulbs: Toward a Theory of Sociotechnical Change*. Cambridge, MA: MIT Press.

BIBLIOGRAPHY

Bijker, W. E., Hughes, T. P. and Pinch, T. (eds) (1987) *The Social Construction of Technological Systems*. Cambridge, MA: MIT Press (selection from Introduction in Scharff and Dusek, pp. 221–32).

Bijker, W. and Law, J. (eds) (1994) *Shaping Technology/Building Reality*. Cambridge, MA: MIT Press.

Blake, W. (1977) *The Complete Poems* (ed. A. Ostriker). London: Penguin.

Bloor, D. (1976) *Knowledge and Social Imagery*. Chicago: University of Chicago Press (2nd edn 1991).

Boardman, P. (1944) *Patrick Geddes: Maker of the Future*. Chapel Hill: University of North Carolina Press.

Borgmann, A. (1984) *Technology and the Character of Contemporary Life*. Chicago: University of Chicago Press (selection in Scharff and Dusek, pp. 293–314).

Borgmann, A. (1995) Information and reality at the turn of the century. *Design Issues*, 11(2), 21–30 (also in Scharff and Dusek, pp. 571–7).

Borgmann, A. (1999) *Holding on to Reality: The Nature of Information at the Turn of the Millennium*. Chicago: University of Chicago Press.

Borgmann, A. (2003) *Power Failure: Christianity and the Culture of Technology*. Grand Rapids, MI: Brazos Press.

Boulding, K. E. (1968) *Beyond Economics: Essays on Society, Religion, and Ethics*. Ann Arbor: University of Michigan Press.

Bowler, P. J. (1992) *The Norton History of the Environmental Sciences*. New York: W. W. Norton.

Bown, N., Burdett, C. and Thurschwell, P. (eds) (2004) *The Victorian Supernatural*. Cambridge: Cambridge University Press.

Braverman, H. (1974) *Labor and Monopoly Capital*. New York: Monthly Review Press.

Brecht, B. (1938) *The Life of Galileo* (trans. D. I. Vesey). London: Methuen.

Breggin, P. and Breggin, G. (1994) *The War against Children: How the Drugs, Programs, and Theories of the Psychiatric Establishment Are Threatening America's Children with a Medical "Cure" for Violence*. New York: St Martin's Press.

Bridges, D. S. (1979) *Constructive Functional Analysis*. London: Pitman.

Bronowski, J. (1965) *William Blake in the Age of Revolution*. New York: Harper & Row.

Bronowski, J. (1973) *The Ascent of Man*. Boston: Little, Brown.

Brouwer, L. E. J. (1907–55) *Collected Works*. Amsterdam: North Holland.

Brown, H. I. (1988) *Rationality*. London: Routledge.

Brumbaugh, R. S. (1966) *Ancient Greek Gadgets and Machines*. New York: Crowell.

Bryld, M. and Lykke, N. (2000) *Cosmodolphins*. London: Zed Books.

Brzezinski, Z. (1970) *Between Two Ages: America's Role in the Technetronic Era*. New York: Viking.

Buchdahl, G. (1961) *The Image of Newton and Locke in the Age of Reason*. London: Sheed and Ward.

Bunge, M. (1967) *Scientific Research II: The Search for Truth*. New York: Springer-Verlag.

BIBLIOGRAPHY

Bunge, M. (1979) Philosophical inputs and outputs of technology. In G. Buliarello and D. B. Doner (eds), *History of Philosophy and Technology*. Urbana: University of Illinois Press (also in Scharff and Dusek, pp. 172–81).

Bunkle, P. (1984) Calling the shots? The politics of Depo-Provera. In R. Arditi, R. D. Klein and S. Minden (eds), *Test Tube Women*. New York: HarperCollins.

Bursill-Hall, G. L. (1972) *Grammatica Speculativa of Thomas of Erfurt*. London: Longman.

Campbell, D. T. (1974) Evolutionary epistemology. In P. A. Schilpp (ed.) *The Philosophy of Karl Popper*. LaSalle, IL: Open Court.

Capra, F. (1982) *The Turning Point: Science, Society and the Rising Culture*. New York: Simon & Schuster.

Carnap, R. (1928) *The Logical Structure of the World* (trans. R. A. George). Berkeley: University of California Press (1967).

Carnap, R., Hahn, H. and Neurath, O. (1929) The scientific conception of the world: the Vienna Circle. In M. Neurath and R. S. Cohen (eds), *Otto Neurath: Empiricism and Sociology*. Dordrecht: Reidel (1973) (selection in Scharff and Dusek, pp. 86–95).

Caro, R. A. (1975) *Power Broker: Robert Moses and the Fall of New York*. New York: Vintage Books.

Carson, R. (1962) *Silent Spring*. Boston: Houghton Mifflin.

Cassirer, E. (1923) 1953 *Philosophy of Symbolic Forms. Volume 1, Mythic Thought* (trans. R. Mannheim). New Haven, CT: Yale University Press (1953).

Cassirer, E. (1948) *The Myth of the State* (trans. R. Mannheim). New Haven, CT: Yale University Press.

Cassirer, E. (1963) *Rousseau, Kant, and Goethe, Two Essays* (trans. J. Gutmann, P. O. Kristeller and J. H. Randall Jr). New York: Harper Books.

Chodorow, N. (1978) *The Reproduction of Mothering*. Berkeley and Los Angeles: University of California Press.

Collins, H. (1985) *Changing Order: Replication and Induction in Scientific Practice*. Chicago: University of Chicago Press.

Commoner, B. (1967) *Science and Survival*. New York: Viking.

Commoner, B. (1975) How poverty causes overpopulation (and not the other way around). *Ramparts*, Aug/Sept, 21–5, 58–9. Also in C. Merchant (ed.), *Ecology*. Atlantic Highlands, NJ: Humanities Press (1994).

Comte, A. (1830) *Introduction to Positive Philosophy* (ed. and trans. F. Ferre). Indianapolis: Hackett Publishing Company (1988) (selection in Scharff and Dusek, pp. 45–59).

Conway, F. and Siegelman, J. (2005) *Dark Hero of the Information Age: In Search of Norbert Wiener, the Father of Cybernetics*. New York: Basic Books.

Cowan, R. S. (1983) *More Work for Mother: The Ironies of Household Technology from the Open Hearth to the Microwave*. New York: Basic Books.

Cowan, R. S. (1997) *A Social History of American Technology*. New York: Oxford University Press.

BIBLIOGRAPHY

Crease, R. P. (ed.) (1997) *Hermeneutics and the Natural Sciences.* Dordrecht: Kluwer Academic.

Cromer, A. (1993) *Uncommon Sense: The Heretical Nature of Science.* New York: Oxford University Press.

Dahrendorf, R. (1965) *Society and Democracy in Germany.* Garden City, NY: Doubleday.

Davidson, B. (1959) *The Lost Cities of Africa.* Boston: Little Brown and Company.

Davidson, B. (1965) *A History of West Africa to the Nineteenth Century.* London: Longmans Green and Company.

Dawkins, R. (1976) *The Selfish Gene.* Oxford: Oxford University Press.

Deleuze, G. (1966) *Bergsonism.* Cambridge, MA: Zone Books (1990).

Deleuze, G. (1988) *The Fold: Leibniz and the Baroque.* Minneapolis: University of Minnesota Press (1993).

Dewey, J. (1931) *Context and Thought.* Berkeley: University of California Press. Also in Dewey, *Later Works, Volume 6.* Carbondale: Southern Illinois University Press.

Diacu, F. and Holmes, P. (1996) *Celestial Encounters: The Origins of Chaos and Stability.* Princeton, NJ: Princeton University Press.

Dobbs, B. J. T. (1991) *The Janus Faces of Genius: The Role of Alchemy in Newton's Thought.* Cambridge: Cambridge University Press.

Dowd, D. (1964) *Thorstein Veblen.* New York: Washington Square Press.

Downs R. E., Kerner, D. and Reyna, S. P. (eds) (1991) *The Political Economy of African Famine.* New York: Gordon and Breach.

Douglas, M. (1980) *Edward Evans-Pritchard.* New York: Viking.

Douglas, M. and Wildavsky, A. (1982) *Risk and Culture: An Essay on the Selection of Technical and Environmental Dangers.* Berkeley: University of California Press.

Dreyfus, H. L. (1965) *Alchemy and Artificial Intelligence.* RAND Corporation paper P-3244.

Dreyfus, H. L. (1972) *What Computers Can't Do: The Limits of Artificial Intelligence.* New York: Harper & Row.

Dreyfus, H. L. (1992) *What Computers Still Can't Do: A Critique of Artificial Reason.* Cambridge, MA: MIT Press.

Dreyfus, H. L. (1999) Anonymity versus commitment: the dangers of education on the Internet. *Ethics and Information Technology,* 1, 15–21 (also in Scharff and Dusek, pp. 578–84).

Dreyfus, H. L. and Dreyfus, S. E. (1986) *Mind over Machine: The Power of Human Intuition and Expertise in the Era of the Computer.* New York: Free Press.

Dreyfus, H. L. and Spinosa, C. (1997) Highway bridges and feasts: Heidegger and Borgmann on how to affirm technology. *Man and World,* 30(2), 159–77 (selection in Scharff and Dusek, pp. 315–26).

Drucker, P. F. (1993) *Post-capitalist Society.* New York: HarperBusiness.

Durkheim, E. (1897) *Suicide* (trans. J. A. Spaulding and G. Simpson). Glencoe, IL: Free Press (1951).

Dusek, V. (1998) Where learned armies clash by night. *Continental Philosophy Review,* 31, 95–106.

BIBLIOGRAPHY

Dusek, V. (1999) *The Holistic Inspirations of Physics: An Underground History of Electromagnetic Theory*. New Brunswick, NJ: Rutgers University Press.

Easlea, B. (1980) *Witch-hunting, Magic and the New Philosophy*. Atlantic Highlands, NJ: Humanities Press.

Eccles, Sir J. C. (1953) *The Neurophysiological Basis of Mind: The Principles of Neurophysiology*. Oxford: Clarendon Press.

Eccles, Sir J. C. (1994) *How the Self Controls Its Brain*. Berlin: Springer-Verlag.

Eddington, Sir A. S. (1934) *New Pathways in Science*. Ann Arbor: University of Michigan Press (1959).

Editors of *Lingua Franca* (2000) *The Sokal Hoax: The Sham that Shook the Academy*. Lincoln: University of Nebraska Press.

Ehrlich, P. R. (1968) *The Population Bomb*. New York: Ballantine.

Ehrlich, P. R. (1986) *The Machinery of Nature*. New York: Simon & Schuster.

Ehrlich, P. R. and Ehrlich, A. H. (1972) *Population, Resources, Environment*. New York: W. H. Freeman.

Ehrlich, P. R. and Ehrlich, A. H. (1991) *The Population Explosion*. New York: Harper-Touchstone.

Ehrlich, P. R. and Feldman, S. S. (1978) *The Race Bomb: Prejudice, Skin Color and Intelligence*. New York: Quadrangle Books.

Elam, M. (1994) Anti anticonstructivism or laying the fears of a Langdon Winner to rest, with Reply by Winner, *Technology and Human Values*, 19(1), 101–9 (selection in Scharff and Dusek, pp. 612–16).

Ellul, J. (1954) *The Technological Society* (trans. J. Wilkinson). New York: Alfred A. Knopf (1964) (selection in Scharff and Dusek, pp. 182–6).

Ellul, J. (1962) *Propaganda: The Formation of Men's Attitudes* (trans. K. Kellen and J. Lerner). New York: Vintage (1973).

Ellul, J. (1980) *The Technological System* (trans. J. Neugroschel). New York: Continuum (selections in Scharff and Dusek, pp. 386–97).

Elsner, H. Jr (1967) *The Technocrats: Prophets of Automation*. New York: Syracuse University Press.

Engels, F. (1882) The part played by labor in the transition from ape to man. Appendix to *Dialectics of Nature* (trans. C. Dutt). New York: International Publishers (1940) (selection in Scharff and Dusek, pp. 71–7).

Engels, F. (1874) On authority. In K. Marx and F. Engels, *Basic Writings on Politics and Philosophy* (ed. L. Feuer). Garden City, NY: Doubleday (1959) (also in Scharff and Dusek, pp. 78–9).

Ernest, P. (1998) *Social Constructivism as a Philosophy of Mathematics*. Albany: State University of New York Press.

Farley, J. and Geison, G. L. (1974) Science, politics, and spontaeous generation in nineteenth century France: the Pasteur–Pouchet debate. *Bulletin of the History of Medicine*, 48, 161–98.

Farrington, B. (1964) *Greek Science: Its Meaning for Us*. Baltimore: Penguin.

BIBLIOGRAPHY

Fausto-Sterling, A. (2000) *Sexing the Body: Gender Politics and the Construction of Sexuality*. New York: Basic Books.

Feenberg, A. (1991) *Critical Theory of Technology*. New York: Oxford University Press.

Feenberg, A. (1992) Subversive rationality: technology, power, and democracy. *Inquiry*, 35(3/4), 301–22 (revised version (2003) Democratic rationalization: technology, power, and freedom, in Scharff and Dusek, pp. 652–65).

Feenberg, A. (1995) *Alternative Modernity: The Technical Turn in Philosophy and Social Theory*. Berkeley: University of California Press.

Feenberg, A. (1999) *Questioning Technology*. London: Routledge.

Feenberg, A. (2002) *Transforming Technology: A Critical Theory Revisited*. New York: Oxford University Press.

Feenberg, A. and Hannay, A. (eds) (1995) *Technology and the Politics of Knowledge*. Bloomington: Indiana University Press.

Feyerabend, P. (1981) Historical introduction. In *Philosophical Papers. Volume 2, Problem of Empiricism*. Cambridge: Cambridge University Press.

Fichte, J. G. (1794) *Theory of Science: Attempt at a Detailed and in the Main Novel Exposition of Logic with Constant Attention to Earlier Authors* (ed. and trans. R. George). Berkeley: University of California Press (1972).

Finley, M. I. (1983a) *Ancient Slavery and Modern Ideology* (ed. B. D. Shaw and R. P. Saller). New York: Penguin.

Finley, M. I. (1983b) *Economy and Society in Ancient Greece*. New York: Penguin.

Firestone, S. (1970) *The Dialectic of Sex: The Case for Feminist Revolution*. New York: Morrow.

Fitzpatrick, J. (1992) The Middle Kingdom, the Middle Sea, and the geographic pivot of history. *Review*, 15(3), 477–521.

Foucault, M. (1976) *History of Sexuality, Volume 1* (trans. R. Hurley). New York: Pantheon Books (1978).

Foucault, M. (1977) *Discipline and Punish: The Birth of the Prison* (trans. A. Sheridan). New York: Pantheon Books.

Fox, N. (2002) *Against the Machine*. Washington, DC: Shearwater Books.

Frank, A. G. (1967) *Capitalism and Underdevelopment in Latin America: Case Studies of Chile and Brazil*. New York: Monthly Review Press.

Frank, A. G. (1998) *Orientation: Global Economy in an Asian Age*. Berkeley: University of California Press.

Fraser, N. (1987) What's critical about critical theory? The case of Habermas and gender. In S. Benhabib and D. Cornell (eds), *Feminism as Critique*. Minneapolis: University of Minnesota Press.

Frazer, Sir J. G. (1890) *The Golden Bough: A Study in Magic and Religion* (3rd edn, 13 vols, 1935). New York: Macmillan.

Freudenthal, G. (1986) *Atomism and Individualism in the Age of Newton*. Dordrecht: Reidel.

Fromm, E. (1961) *Marx's Concept of Man, with a Translation of Marx's Economic and Philosophical Manuscripts by T. B. Bottomore*. New York: Frederick Ungar.

BIBLIOGRAPHY

Fuller, S. (1988) *Social Epistemology*. Bloomington: Indiana University Press.

Fuller, S. (1997) *Science*. Minneapolis: University of Minnesota Press.

Galbraith, J. K. (1967) *The New Industrial State*. New York: New American Library.

Galison, P. (1987) *How Experiments End*. Chicago: University of Chicago Press.

Gasman, D. (1971) *The Scientific Origins of National Socialism: Social Darwinism in Ernst Haeckel and the German Monist League*. New York: American Elsevier.

Geison, G. L. (1995) *The Private Science of Louis Pasteur*. Princeton, NJ: Princeton University Press.

Gendron, B. (1977) *Technology and the Human Condition*. New York: St Martin's Press.

Ghiselin, M. T. (1969) *The Triumph of the Darwinian Method*. Berkeley: University of California Press.

Gilbert, W. (1996) A vision of the Grail. In D. Kevles and L. Hood (eds), *The Code of Codes*. Cambridge, MA: Harvard University Press.

Glendinning, C. (1990) Notes toward a neo-Luddite manifesto. *Utne Reader*, 38(1), 50–3 (also in Scharff and Dusek, pp. 603–5).

Golinski, J. (1998) *Making Natural Knowledge*. Cambridge: Cambridge University Press.

González, R. J. (2001) *Zapotec Science: Farming and Food in the Northern Sierra of Oaxaca*. Austin: University of Texas Press.

Goodyear, C. (1855) *Gum Elastic*. New Haven, CT: published by author. Reproduced in *A Centennial Volume of the Writings of Charles Goodyear and Thomas Hancock*. Boston: Centennial Committee, American Chemical Society (1939).

Goodwin, F. (1992) Violence initiative. Address to the Meeting of the National Mental Health Advisory Council, February 11.

Gould, S. J. (1980) Darwin's middle way. In *The Panda's Thumb*. New York: Norton.

Graham, A. C. (1978) *Later Mohist Logic, Ethics, and Science*. Hong Kong: Chinese University Press (reissued New York: Columbia University Press, 2003).

Grandy, R. E. (1977) *Advanced Logic for Applications*. Boston: Reidel.

Greenfield, P. (1991) Language, tool and brain: phylogeny and ontogeny of hierarchically organized sequential behavior. *Behavior and Brain Sciences*, 14, 531–95.

Gurwitsch, A. (1964) *Field of Consciousness*. Pittsburgh: Duquesne University Press.

Habermas, J. (1970) *Toward a Rational Society*. Boston: Beacon Press (selection in Scharff and Dusek, pp. 530–5).

Habermas, J. (1971) *Knowledge and Human Interests* (trans. J. Shapiro). London: Heinemann.

Habermas, J. (1987) *Theory of Communicative Action, Volume 2. Lifeworld and System: A Critique of Functionalist Reason* (trans. T. McCarthy). Boston: Beacon Press.

Habib, I. (1969) Potentialities of capitalist development in the economy of Mughal India. *Journal of Economic History*, 29(1), 32–78.

Hacking, I. (1983) *Representing and Intervening*. Cambridge: Cambridge University Press.

Hacking, I. (1999) *The Social Construction of What?* Cambridge, MA: Harvard University Press.

BIBLIOGRAPHY

Haeckel, E. (1868) *The History of Creation*. New York: D. Appleton.

Haeckel, E. (1869) *Ueber Arbeitsheilung in Natur und Menschenleben*. Berlin: n.p.

Hammer, C. and Dusek, V. (1995) Brain difference research and learning styles literature: from equity to discrimination. *The Feminist Teacher*, Fall/Winter, 76–83.

Hammer, C. and Dusek, R. V. (1996) Anthropological stories and educational results [peer commentary]. *Behavior and Brain Science*, June, 357.

Hannah-Barbera (2003) *Scooby-Doo: Space Ape at the Cape* (videotape). Haywood, CA: Warner Brothers.

Hanson, N. R. (1958) *Patterns of Discovery*. Cambridge: Cambridge University Press.

Hanson, N. R. (1964) Stability proofs and consistency proofs: a loose analogy. *Philosophy of Science*, 31(4), 301–18.

Haraway, D. (1985) Manifesto for cyborgs: science, technology and socialist feminism in the 1980s. *Socialist Review*, 80, 65–108 (expanded version in *Simians, Cyborgs and Women*, selection of latter version in Scharff and Dusek, pp. 429–50).

Haraway, D. (1989) *Primate Visions: Gender, Race, and Nature in the World of Modern Science*. New York: Routledge.

Haraway, D. (1991) *Simians, Cyborgs, and Women: The Reinvention of Nature*. New York: Routledge.

Hardin, G. (1972) *Exploring New Ethics for Survival: The Voyage of the Spaceship Beagle*. New York: Viking.

Hardin, G. (1980) *Promethean Ethics: Living with Death, Competition, and Triage*. Seattle: University of Washington Press.

Harding, S. (ed.) (1976) *Can Theories Be Refuted? Essays on the Duhem–Quine Thesis*. Boston: Reidel.

Harding, S. (1986) *The Science Question in Feminism*. Ithaca, NY: Cornell University Press.

Harding, S. (1991) *Whose Science, Whose Knowledge?* Ithaca, NY: Cornell University Press.

Harding, S. (1998) *Is Science Multicultural? Postcolonialisms, Feminisms and Epistemologies*. Bloomington: Indiana University Press (selection in Scharff and Dusek, pp. 154–69).

Harré, R. (1970) *The Principles of Scientific Thinking*. Chicago: University of Chicago Press.

Harrington, A. (1996) *Reenchanted Science*. Cambridge, MA: Harvard University Press.

Havelock, E. A. (1963) *Preface to Plato*. Cambridge, MA: Belknap Press.

Hayek, F. A. (1952) *The Sensory Order: An Inquiry into the Foundations of Theoretical Psychology*. Chicago: University of Chicago Press.

Hayek, F. A. (1955) *The Counter-revolution in Science: Studies in the Abuse of Reason*. Glencoe, IL: Free Press.

Hayles, N. K. (1990) *Chaos Bound: Orderly Disorder in Contemporary Literature and Science*. Ithaca, NY: Cornell University Press.

BIBLIOGRAPHY

Hayles, N. K. (ed.) (1991) *Chaos and Order: Complex Dynamics in Literature and Science*. Chicago: University of Chicago Press.

Hearnshaw, L. S. (1979) *Cyril Burt: Psychologist*. London: Hodder and Stoughton.

Hebb, D. O. (1949) *The Organization of Behavior*. New York: Wiley.

Heelan, P. J. (1983) *Space Perception and the Philosophy of Science*. Berkeley: University of California Press.

Heer, F. (1974) *The Challenge of Youth* (trans. G. Skelton). Montgomery: University of Alabama Press.

Hegel, G. W. F. (1807) *The Phenomenology of Mind* (trans. J. B. Baillie). New York: Harper & Row (1967).

Hegel, G. W. F. (1812–16) *Logic* (trans. A. V. Miller). Oxford: Oxford University Press (1969).

Heidegger, M. (1916) *Die Kategorien- und Bedeutungslehre des Duns Scotus*. Tübingen: Nachdruck (1972).

Heidegger, M. (1927) *Being and Time* (trans. J. Macquarrie and E. Robinson). Oxford: Blackwell (1962).

Heidegger, M. (1954) The question concerning technology. In *The Question Concerning Technology and Other Essays* (trans. W. Lovitt). New York: Harper & Row (1977) (also in Scharff and Dusek, pp. 252–65).

Heilbroner, R. L. (1953) *The Worldly Philosophers: The Lives, Times, and Ideas of the Great Economic Thinkers*. New York: Simon & Schuster.

Heilbroner, R. L. (1967) Do machines make history? *Technology and Culture*, 8, 335–45 (also in Scharff and Dusek, pp. 398–404).

Heilbroner, R. L. (1978) Inescapable Marx. *New York Review of Books*, 25(11).

Heims, S. J. (1980) *John von Neumann and Norbert Wiener: From Mathematics to the Technologies of Life and Death*. Cambridge, MA: MIT Press.

Heims, S. J. (1991) *Constructing a Science for Postwar America: The Cybernetics Group 1946–1953*. Cambridge, MA: MIT Press.

Heisenberg, W. (1958) *Physics and Philosophy: The Revolution in Modern Science*. New York: Harper.

Heisenberg, W. (1971) *Physics and Beyond: Encounters and Conversations* (trans. A. J. Pomerans). New York: Harper.

Heisler, R. (1989) Michael Maier and England. *The Hermetic Journal* (www.levity.com/alchemy/h_maier.html).

Heldke, L. (1988) John Dewey and Evelyn Fox Keller: a shared epistemological tradition. *Hypatia*, 3(summer), 114–44.

Henderson, L. D. (1983) *The Fourth Dimension and Non-Euclidean Geometry in Modern Art*. Princeton, NJ: Princeton University Press.

Henderson, L. D. (1998) *Duchamp in Context: Science and Technology in the Large Glass and Related Works*. Princeton, NJ: Princeton University Press.

Herf, J. (1984) *Reactionary Modernism: Technology, Culture, and Politics in Weimar and the Third Reich*. Cambridge: Cambridge University Press.

BIBLIOGRAPHY

Herken, G. (2002) *Brotherhood of the Bomb: The Tangled Lives and Loyalties of Robert Oppenheimer, Ernest Lawrence, and Edward Teller*. New York: Henry Holt and Co.

Hertsgaard, M. (1997) Our real China problem. *Atlantic Monthly*, November.

Hesse, M. (1966) *Models and Analogies in Science*. Evanston, IL: Northwestern University Press.

Heyl, B. S. (1968) The Harvard Pareto Circle. *Journal of the History of the Behavioral Sciences*, 4(4), 316–34.

Hobsbawm, E. (1962) *The Age of Revolution*. New York: New American Library.

Horton, R. (1967) African traditional thought and Western science. In B. Wilson (ed.), *Rationality*. Oxford: Blackwell (1970).

Hubbard, R. (1983) Have only men evolved? In S. Harding and M. B. Hintikka (eds), *Discovering Reality: Feminist Perspectives on Epistemology, Metaphysics, Methodology, and Philosophy of Science*. Dordrecht: Reidel.

Hughes, T. P. (2004) *The Human-built World: How to Think about Technology and Culture*. Chicago: University of Chicago Press.

Husserl, E. (1936) *The Crisis of European Science and Transcendental Phenomenology: An Introduction to Phenomenological Philosophy*. Evanston, IL: Northwestern University Press (1970).

Ihde, D. (1990) *Technology and the Lifeworld: From Garden to Earth*. Bloomington: Indiana University Press (selection in Scharff and Dusek, pp. 507–29).

Ihde, D. (1991) *Instrumental Realism: The Interface between Philosophy of Science and Philosophy of Technology*. Bloomington: Indiana University Press.

Ihde, D. (1998) *Expanding Hermeneutics: Visualism in Science*. Evanston, IL: Northwestern University Press.

Ihde, D. and Selinger, E. (eds) (2003) *Chasing Technoscience: Matrix for Materiality*. Bloomington: Indiana University Press.

Jarvie, I. C. (1967) Technology and the structure of knowledge. *Dimensions of Exploration* pamphlet, Department of Industrial Arts and Technology, College of Arts and Sciences, Oswego, NY. Revised reprint in C. Mitcham and R. Mackey (eds), *Philosophy and Technology*. New York: Free Press.

Jay, M. (1993) *Downcast Eyes: The Degradation of Sight in Twentieth-century French Thought*. Berkeley: University of California Press.

Jeansonne, G. (1974) The automobile and American morality. *Journal of Popular Culture*, 8, 125–31. Reprinted in L. Hickman and A. Al-Hibri (eds), *Technology and Human Affairs*. St Louis, MO: C. V. Mosby.

Joseph, G. G. (1991) *The Crest of the Peacock: Non-European Roots of Mathematics*. New York: Penguin.

Kahneman, D. and Tversky, A. (1973) On the psychology of prediction. *Psychological Review*, 80, 237–51.

Kant, I. (1781) *The Critique of Pure Reason* (trans. W. S. Pluhar). Indianapolis: Hackett (1996).

BIBLIOGRAPHY

Kant, I. (1791) *The Critique of Judgment* (trans. W. S. Pluhar). Indianapolis: Hackett (1987).

Kaplan, F. (1983) *The Wizards of Armageddon*. New York: Simon & Schuster.

Keller, E. F. (1985) *Reflections on Gender and Science*. New Haven, CT: Yale University Press.

Kevles, D. J. (1977) *The Physicists: The History of a Scientific Community in Modern America*. New York: Knopf.

Kinder, H. and Hilgemann, W. (1964) *The Anchor Atlas of World History, Volume 1* (trans. E. A. Menze). New York: Bantam Doubleday Dell (1974).

Kitcher, P. (1993) *The Advancement of Science: Science without Legend, Objectivity without Illusions*. Oxford: Oxford University Press.

Kline, S. J. (1985) What is technology? *Bulletin of Science, Technology and Society*, 1, 215–18 (also in Scharff and Dusek, pp. 210–12).

Koertge, N. (ed.) (1997) *A House Built on Sand: Flaws in the Cultural Studies Account of Science*. Oxford: Oxford University Press.

Kowarski, L. (1971) Scientists as magicians: since 1945. Boston Colloquium for the Philosophy of Science, October 26.

Kuhn, T. (1962) *The Structure of Scientific Revolutions*. Chicago: University of Chicago Press (2nd edn 1970).

Labib, S. Y. (1969) Capitalism in medieval Islam. *Journal of Economic History*, 29(1), 79–96.

Landau, M. (1991) *Narratives of Human Evolution*. New Haven, CT: Yale University Press.

Landes, D. S. (1983) *Revolution in Time: Clocks and the Making of the Modern World*. Cambridge, MA: Belknap Press.

Lane, F. G. (1969) Meanings of capitalism. *Journal of Economic History*, 29(1), 5–12.

Lapp, R. (1965) *The New Priesthood: The Scientific Elite and the Uses of Power*. New York: Harper & Row.

Lappé, F. M. and Collins, J., with Fowler, C. (1979) *Food First: Beyond the Myth of Scarcity*. New York: Ballantine.

Lappé, F. M. and Collins, J. (1982) *World Hunger: Ten Myths*, 4th edn. San Francisco: Institute for Food and Development Policy.

Laplace, P. S. de (1813) *Philosophical Essay on Probabilities* (trans. F. W. Truescott and F. L. Emory). New York: Dover Publications (1951).

Latour, B. (1987) *Science in Action: How to Follow Scientists and Engineers through Society*. Cambridge, MA: Harvard University Press.

Latour, B. (1988) A relativist account of Einstein's relativity. *Social Studies of Science*, 18, 3–44.

Latour, B. (1992) One more turn after the social turn . . . In E. McMullin (ed.), *The Social Dimensions of Science*. Notre Dame, IN: University of Notre Dame Press.

Latour, B. (1993) *We Have Never Been Modern* (trans. C. Porter). Cambridge, MA: Harvard University Press.

BIBLIOGRAPHY

Latour, B. (1996) Do scientific objects have a history? Pasteur and Whitehead in a bath of lactic acid. *Common Knowledge*, 5(1), 76–91.

Latour, B. (1999) *Pandora's Hope*. Cambridge, MA: Harvard University Press (selection in Scharff and Dusek, pp. 126–37).

Latour, B. and Woolgar, S. (1979) *Laboratory Life: The Construction of Scientific Facts*. London: Sage (with a new postscript, Princeton, NJ: Princeton University Press, 1986).

Leibniz, G. F. von (1714) *Discourse on Metaphysics, Correspondence with Arnauld, and Monadology* (trans. G. R. Montgomery). La Salle, IL: Open Court (1902).

Lettvin, J., Maturana, H. R., McCulloch, W. S. and Pitts, W. H. (1959) What the frog's eye tells the frog's brain. *Proceedings of the IRE*, 47(11), 1940–59 (reprinted in W. S. McCulloch, *Embodiments of Mind*. Cambridge, MA: MIT Press).

Lévy-Bruhl, L. (1910) *Primitive Mentality*. Boston: Beacon Press (1966).

Lévy-Bruhl, L. (1949) *Notebooks on Primitive Mentality*. Oxford: Blackwell (1973).

Locke, J. (1689) *An Essay Concerning Human Understanding* (ed. P. Nidditch). Oxford: Oxford University Press (1975).

Lovelock, J. E. (1984) *Gaia: A New Look at Life on Earth*. Oxford: Oxford University Press.

Lowrance, W. W. (1976) *Of Acceptable Risk: Science and the Determination of Safety*. Los Altos, CA: William Kaufmann.

Lukács, G. (1923) *History and Class Consciousness*. Cambridge, MA: MIT Press (1971).

Lyotard, J.-F. (1979) *The Post-modern Condition*. Minneapolis: University of Minnesota Press (1984).

McCulloch, W. S. (1943–64) *Embodiments of Mind*. Cambridge, MA: MIT Press (1988).

McDermott, R. E., Mikulak, R. J. and Beauregard, M. R. (1996) *The Basics of FMEA*. Quality Resources.

McLuhan, M. (1964) *Understanding Media: Extensions of Man*. New York: McGraw-Hill.

Macpherson, C. B. (1962) *The Political Theory of Possessive Individualism: Hobbes to Locke*. Oxford: Oxford University Press.

Malinowski, B. (1922) *Argonauts of the Western Pacific: An Account of Native Enterprise and Adventure in the Archipelagoes of Melanesian New Guinea*. New York: Dutton (1961).

Malinowski, B. (1925) Magic, science, and religion. In *Magic, Science, and Religion and Other Essays*. New York: Doubleday Anchor (1948).

Malthus, T. (1803) *An Essay on the Principle of Population*. New York: Penguin (1983).

Manuel, F. E. (1962) *The Prophets of Paris*. Cambridge, MA: Harvard University Press.

Mannheim, K. (1929) *Ideology and Utopia* (trans. L. Wirth and E. Shils). London: Routledge & Kegan Paul (1936).

Mannheim, K. (1935) *Man and Society in an Age of Reconstruction*. London: Routledge.

Mannheim, K. (1950) *Freedom, Power, and Democratic Planning*. New York: Oxford University Press.

BIBLIOGRAPHY

Marcuse, H. (1932) The foundations of historical materialism. In *Studies in Critical Philosophy*. Boston: Beacon Press (1972).

Marcuse, H. (1964) *One-dimensional Man*. Boston: Beacon Press (selection in Scharff and Dusek, pp. 405–12).

Marcuse, H. (1965) Industrialism and capitalism in the work of Max Weber. In *Negations* (trans. J. Shapiro). Boston: Beacon Press (1968).

Marglin, S. (1974) What do bosses do? *Review of Radical Political Economy*, Summer, 33–60.

Marx, K. (1852) *The Eighteenth Brumaire of Louis Napoleon*. New York: International Publishers.

Marx, K. (1859) *A Contribution to the Critique of Political Economy*. New York: International Publishers (1970) (selection in Scharff and Dusek, pp. 69–71).

Marx, K. (1867–87) *Capital*, three volumes. New York: Penguin (1992–3).

Marx, K. (1963) *Early Writings* (ed. T. Bottomore). New York: McGraw-Hill.

Marx, K. and Engels, F. (1846) *German Ideology*. Moscow: Progress Publishers (1968).

Marx, K. and Engels, F. (1848) *The Communist Manifesto*, and *The Principles of Communism*, by F. Engels (trans. P. Sweezey). New York: Monthly Review Press.

Marx, K. and Engels, F. (1954) *Marx and Engels on Malthus: Selections from the Writings of Marx and Engels Dealing with the Theories of Thomas Robert Malthus* (trans. D. L. Meek and R. L. Meek). New York: International Publishers.

Mathews, F. (1991) *The Ecological Self*. London: Routledge.

Mathews, F. (2003) *For Love of Matter: A Contemporary Panpsychism*. Albany: State University of New York Press.

Mayo, D. and Hollander, R. (1991) *Acceptable Evidence: Science and Values in Risk Management*. New York: Oxford University Press.

Mayr, E. (1957) Species concepts and definitions. In E. Mayr (ed.), *The Species Problem*. Washington, DC: AAAS.

Mead, G. H. (1932) *The Philosophy of the Present* (ed. A. E. Murphy). Chicago: Open Court.

Merchant, C. (1980) *The Death of Nature: Women, Society and the Scientific Revolution*. New York: Harper & Row.

Merchant, C. (1983) Mining the earth's womb. In J. Rothschild (ed.), *Feminist Perspectives on Technology*. Oxford: Pergamon Press (also in Scharff and Dusek, pp. 417–28).

Merleau-Ponty, M. (1942) *The Structure of Behavior* (trans. A. L. Fisher). Boston: Beacon Press (1963).

Merleau-Ponty, M. (1945) *Phenomenology of Perception* (trans. C. Smith). New York: Humanities Press (1962).

Merleau-Ponty, M. (1964a) *Signs*. Evanston, IL: Northwestern University Press.

Merleau-Ponty, M. (1964b) *The Visible and the Invisible; Followed by Working Notes* (ed. C. Lefort, trans. A. Lingis). Evanston IL: Northwestern University Press (1968).

BIBLIOGRAPHY

Merton, R. K. (1938) Science and the social order. *Philosophy of Science*, 5(3), 321–37 (also in *Social Theory and Social Structure*).

Merton, R. K. (1942) Science and democratic structure. *Journal of Legal and Political Sociology*, 1 (also in *Social Theory and Social Structure*).

Merton, R. K. (1947) *Social Theory and Social Structure*. Glencoe, IL: Free Press.

Merton, R. K. (1961) Singletons and multiples in scientific discovery. *Proceedings of the American Philosophical Society*, 105(5), 470–86; reprinted in *Sociology of Science*.

Merton, R. K. (1973) *Sociology of Science*. Chicago: University of Chicago Press.

Mill, J. S. (1843) *A System of Logic*. Charlottesville, VA: Lincoln-Rembrandt.

Mills, C. W. (1962) *The Marxists*. New York: Dell.

Minsky, M. L. and Papert, S. (1969) *Perceptrons: Introduction to Computational Geometry*. Cambridge, MA: MIT Press (exp. edn 1990).

Moir, A. and Jessel, D. (1992) *Brain Sex: The Real Difference between Men and Women*. New York: Dell.

Moss, L. (2002) *What Genes Can't Do*. Cambridge, MA: MIT Press.

Moss, L. (2004a) Human nature, Habermas, and the anthropological framework of critical theory. Unpublished.

Moss, L. (2004b) Human nature, the genetic fallacy, and the philosophy of anthropogenesis. Unpublished.

Mumford, L. (1934) *Technics and Civilization*. New York: Harcourt Brace Jovanovich (1963).

Mumford, L. (1966) The concept of the megamachine. In P. H. Oehser (ed.), *Knowledge among Men: Eleven Essays on Science Culture and Society Commemorating the 200th Anniversary of the Birth of James Smithson*. New York: Simon & Schuster (also in Scharff and Dusek, pp. 348–51).

Mumford, L. (1967) *The Myth of the Machine: Technics and Human Development*. New York: Harcourt Brace Jovanovich (selection in Scharff and Dusek, pp. 344–8).

Murray, G. (1925) *Five Stages of Greek Religion*, 2nd edn. New York: Columbia University Press.

Myrdal, G. (1942) *An American Dilemma: The Negro Problem and Modern Democracy*. New York: Harper & Row.

Myrdal, G. (1960) *Beyond the Welfare State: Economic Planning and Its International Implications*. New Haven, CT: Yale University Press.

Nabi, I. (pseudonym of R. Levins and R. Lewontin) (1981) On the tendencies of motion. *Science and Nature*, 4, 62–6.

Naess, A. (1973) The shallow and the deep, long range ecology movement. *Inquiry*, 18, 95–100 (also in Scharff and Dusek, pp. 367–70).

Needham, J. (1954–) *Science and Civilisation in China*. Cambridge: Cambridge University Press.

Nelson, D., Joseph, G. G. and Williams, J. (1993) *Multicultural Mathematics: Teaching Mathematics from a Global Perspective*. Oxford: Oxford University Press.

BIBLIOGRAPHY

Nelson, L. H. (1990) *Who Knows: From Quine to Feminist Empricism*. Philadelphia: Temple University Press.

Noble, D. (1984) *The Forces of Production: A Social History of Industrial Automation*. Oxford: Oxford University Press.

Noble, D. (1992) *A World without Women: The Christian Clerical Culture of Western Science*. New York: Oxford University Press.

Noble, D. (1993) Upon opening the black box and finding it empty: social constructivism and the philosophy of technology. *Science, Technology and Human Values*, 18(3), 362–78 (also in Scharff and Dusek, pp. 233–44).

Norris, C. (1992) *Uncritical Theory: Postmodernism, Intellectuals and the Gulf War*. Amherst: University of Massachusetts Press.

Oakley, A. (1974) *The Sociology of Housework*. New York: Pantheon Books.

Odum, H. T. (1970) *Environment, Power, and Society*. New York: Wiley-Interscience.

Ogburn, W. F. (1922) *Social Change with Respect to Culture and Original Nature*. New York: B. W. Huebsch.

Ong, W. J. (1958) *Ramus: Method, and the Decay of Dialogue: from the Art of Discourse to the Art of Reason*. Cambridge, MA: Harvard University Press.

Ormrod, S. (1994) Let's nuke the dinner: discursive practices of gender in the creation of new cooking process. In C. Cokburn and R. F. Dilic (eds), *Bringing Technology Home: Gender and Technology in a Changing Europe*. Buckingham: Open University Press.

Ortega y Gasset, J. (1939) *History as a System, and Other Essays on Philosophy of History*. New York: Norton (1962).

Owen, A. (2004) *The Place of Enchantment: British Occultism and the Culture of the Modern*. Chicago: University of Chicago Press.

Oxford University Press (1995–6) Books new and forthcoming. Flyer.

Pacey, A. (1983) *The Culture of Technology*. Cambridge, MA: MIT Press.

Pacey, A. (1990) *Technology in World Civilization*. Cambridge, MA: MIT Press.

Peirce, C. S. (1869) Ockam, lecture 2 and author's preamble. In *Writings of Charles S. Peirce: A Chronological Edition, Volume 2, 1867–1871* (ed. E. C. Moore). Bloomington: Indiana University Press.

Penfield, W. (1975) *The Mystery of the Mind: A Critical Study of Consciousness and the Human Brain*. Princeton, NJ: Princeton University Press.

Penrose, R. (1994) *Shadows of the Mind: A Search for the Missing Science of Consciousness*. Oxford: Oxford University Press.

Perlin, F. (1983) Proto-industrialization and pre-colonial south Asia. *Past and Present*, 98(Feb), 30–95.

Perrin, N. (1979) *Giving up the Gun: Japan's Reversion to the Sword, 1543–1879*. Boston: David R. Godine.

Petchesky, R. P. (1987) Fetal images: the power of visual culture in the politics of reproduction. In M. Stanhope (ed.), *Reproductive Technologies: Gender, Motherhood and Medicine*. Minneapolis: University of Minnesota Press.

BIBLIOGRAPHY

Perrow, C. (1984) *Normal Accidents: Living with High-risk Technologies*. New York: Basic Books.

Piaget, J. (1930) *The Child's Conception of Physical Causality* (trans. M. Gabain). Paterson, NJ: Littlefield, Adams (1960).

Piaget, J. (1952) *The Child's Conception of Number* (trans. C. Gattegno and F. M. Hodgson). London: Routledge & Paul.

Pickering, A. (1995) *The Mangle of Practice: Time, Agency and Science*. Chicago: University of Chicago Press.

Pitt, J. (2000) *Thinking about Technology: Foundations of Philosophy of Technology*. New York: Seven Bridges Press.

Plato (1992) *Republic* (trans. G. M. A. Grube and C. D. C. Reeve). Indianapolis: Hackett (selection in Scharff and Dusek, pp. 8–18).

Poincaré, H. (1902) *Science and Hypothesis* (trans. R. B. Halstead). New York: Dover Publications.

Poincaré, H. (1913) *Last Thoughts*. New York: Dover Publications.

Polanyi, M. (1958) *Personal Knowledge: Towards a Post-critical Philosophy*. Chicago: University of Chicago Press.

Popper, K. (1934) *The Logic of Scientific Discovery*. New York: Basic Books (1959).

Popper, K. (1945) *The Open Society and Its Enemies*. London: Routledge.

Popper, K. (1962) *Conjectures and Refutations: The Growth of Scientific Knowledge*. New York: Basic Books.

Poundstone, W. (1992) *Prisoner's Dilemma: John von Neumann, Game Theory, and the Puzzle of the Bomb*. New York: Doubleday.

Proctor, R. N. (1999) *The Nazi War on Cancer*. Princeton, NJ: Princeton University Press.

Provine, W. B. (1986) *Sewall Wright and Evolutionary Biology*. Chicago: University of Chicago Press.

Putnam, H. (1981) *Reason, Truth and History*. Cambridge: Cambridge University Press.

Quammen, D. (1996) *The Song of the Dodo: Island Biogeography in an Age of Extinctions*. New York: Simon & Schuster.

Quine, W. v. O. (1951) Two dogmas of empiricism. In *From a Logical Point of View*. Cambridge, MA: Harvard University Press (1961).

Quinton, A. (1967) Cut-rate salvation. *New York Review of Books*, 9(9), Nov. 23.

Ravetz, J. R. (1971) *Scientific Knowledge and Its Social Problems*. New York: Oxford University Press.

Reid, C. (1970) *Hilbert*. New York: Springer-Verlag.

Reisch, G. (2005) *How the Cold War Transformed Philosophy of Science*. Cambridge: Cambridge University Press.

Rescher, N. (1983) *Risk: A Philosophical Introduction to the Theory of Risk Evaluation and Management*. Lanham, MD: University Press of America.

Reyna, S. P. and Downs, R. E. (1999) *Deadly Developments: Capitalism, States and War*. Amsterdam: Gordon and Breach.

BIBLIOGRAPHY

Richards, J. F. (1990) The seventeenth-century crisis in south Asia. *Modern Asian Studies*, 24(4), 625–38.

Rifkin, J. (1983) *Algeny*. New York: Viking Press.

Roberts, N. H. (1987) *Fault Tree Handbook*. Washington, DC: US Government Printing Office.

Robins, K. and Webster, F. (1990) Athens without slaves . . . or slaves without Athens? The neurosis of technology. *Science as Culture*, 7–53.

Rodinson, M. (1974) *Islam and Capitalism* (trans. B. Pearce). New York: Pantheon Books.

Ronan, C. (1978–) *The Shorter Science and Civilization in China*. Cambridge: Cambridge University Press.

Rosenblatt, M. (ed.) (1984) *Errett Bishop: Reflections on Him and His Research*. Providence, RI: American Mathematical Society.

Ross, A. (1991) *Strange Weather: Culture, Science, and Technology in the Age of Limits*. London: Verso.

Ross, A. (1996) Introduction. In *Science Wars*. Durham, NC: Duke University Press.

Ross, A. (1998) *Real Love: In Pursuit of Cultural Justice*. New York: New York University Press.

Rossi, P. (1970) *Philosophy, Technology, and the Arts in the Early Modern Era*. New York: Harper & Row.

Rothman, B. K. (1986) *The Tentative Pregnancy: Prenatal Diagnosis and the Future of Motherhood*. New York: Viking.

Rothman, B. K. (2001) *The Book of Life: A Personal and Ethical Guide to Race, Normality, and the Implications of the Human Genome Project*. Boston: Beacon Press.

Rothschild, J. (ed.) (1983) *Machina ex Dea: Feminist Perspectives on Technology*. Oxford: Pergamon Press.

Rousseau, J.-J. (1750) *Discourse on the Science and the Arts*. In *The First and Second Discourses* (ed. R. D. Masters, trans. R. D. and J. R. Masters). New York: St Martin's Press (1964) (selection in Scharff and Dusek, pp. 60–5).

Rousseau, J.-J. (1761) *Julie; or, The New Eloise: Letters of Two Lovers, Inhabitants of a Small Town at the Foot of the Alps* (trans. J. H. McDowell). University Park: Pennsylvania State University Press (1968).

Rousseau, J.-J. (1762) *Emile: Or, on Education* (trans. A. Bloom). New York: Basic Books (1979).

Russell, B. (1914) *Our Knowledge of the External World*. New York: New American Library of World Literature (1958).

Sahlins, M. (1976) *Culture and Practical Reason*. Chicago: University of Chicago Press.

Saint-Simon, H. C. de (1952) *Selected Writings* (ed. and trans. F. Markham). Oxford: Blackwell.

Sale, K. (1995) *Rebels against the Future*. Reading, MA: Addison Wesley.

Salleh, A. (1984) Deeper than deep ecology: the eco-feminist connection. *Environmental Ethics*, 6(4), 339–45.

BIBLIOGRAPHY

Saxton, M. (1984) Born and unborn: implications of the reproductive technologies for people with disabilities. In R. Arditti, R. D. Klein and S. Mindin (eds), *Test-tube Woman: What Future for Motherhood?* Boston: Routledge.

Saxton, M. (1998) Disability rights and selective abortion. In R. Solinger (ed.), *Abortion Wars: A Half Century of Struggle, 1950–2000.* Berkeley: University of California.

Scharff, R. S. and Dusek, V. (eds) (2002) *Philosophy of Technology: The Technological Condition, an Anthology.* Oxford: Blackwell.

Schivelbusch, W. (1979) *The Railroad Journey: The Industrialization of Time.* New York: Urizen. Reissued as *The Railway Journey: The Industrialization of Time and Space in the 19th Century.* Berkeley: University of California Press (1986).

Schumpeter, J. A. (1950) *Capitalism, Socialism, and Democracy.* New York: Harper & Row.

Scott, H. (1974) *Does Socialism Liberate Women? Experiences from Eastern Europe.* Boston: Beacon Press.

Searle, J. (1995) *The Construction of Social Reality.* New York: Free Press.

Shapin, S. (1994) *A Social History of Truth.* Chicago: University of Chicago Press.

Shapin, S. and Schaffer, S. (1985) *The Leviathan and the Air Pump: Hobbes, Boyle and the Experimental Life.* Princeton, NJ: Princeton University Press.

Sherman, D. H. (1998) Cover story. *Lingua Franca*, September, 24–6.

Simmel, G. (1900) *The Philosophy of Money.* London: Routledge & Kegan Paul (1978).

Sivin, N. (1973) Copernicus in China. *Studia Copernicana* (Warsaw), 6, 63–122. Reprinted in *Science in Ancient China. Researches and Reflections.* Aldershot: Variorum (1995).

Sivin, N. (1986) On the limits of emprical knowledge in Chinese and Western science. In J. T. Fraser et al. (eds), *Time, Science and Society in China and the West.* Amherst: University of Massachusetts Press. Expanded version in *Medicine, Philosophy, and Religion in Ancient China.* Brookfield, VT: Ashgate, Variorum (1995).

Sivin, N. (1984) Reflections on "Nature on trial." In R. S. Cohen and M. Wartofsky (eds), *Methodology, Metaphysics and the History of Science. In Memory of Benjamin Nelson.* Dordrecht: Reidel.

Skinner, B. F. (1966) *Walden II.* New York: Macmillan.

Skinner, B. F. (1971) *Beyond Freedom and Dignity.* New York: Knopf.

Slovic, P., Fischhoff, B. and Lichtenstein, S. (1981) Perceived risk, psychological factors and social implications. *Proceedings of the Royal Society of London*, A376, 17–34.

Smeds, R., Huida, O., Haavio-Mannila, E. and Kauppinen-Toropainen, K. (1994) Sweeping away the dust of tradition: vacuum cleaning as a site of technical and social innovation. In C. Cokburn and R. F. Dilic (eds), *Bringing Technology Home: Gender and Technology in a Changing Europe.* Buckingham: Open University Press.

Smith, A. (1776) *An Inquiry into the Nature and Causes of the Wealth of Nations.* London: Methuen (1904).

Smuts, J. C. (1926) *Holism and Evolution.* New York: Macmillan.

BIBLIOGRAPHY

Soble, A. (1995) In defense of Bacon. *Canadian Journal of Philosophy, Philosophy of the Social Sciences*, 25(2), 192–215. Revised version in N. Koertge (ed.), *A House Built on Sand: Flaws in the Cultural Studies Account of Science*. New York: Oxford University Press (1997) (also in Scharff and Dusek, pp. 451–67).

Sokal, A. (1996) Transgressing the boundaries: toward a transformative hermeneutics of quantum gravity. *Social Text*, 46/47, 217–52.

Spence, J. D. (1984) *The Memory Palace of Matteo Ricci*. New York: Viking Penguin.

Sperry, R. (1983) *Science and Moral Priority: Merging Mind, Brain, and Human Values*. New York: Columbia University Press.

Srole, C. (1987) "A blessing to mankind, and especially womenkind": the typewriter and the feminization of clerical work, Boston, 1869–1920. In B. D. Wright (ed.), *Women, Work, and Technology: Transformations*. Ann Arbor: University of Michigan Press.

Stadler, F. (1982) *Arbeiterbildung in der Zwischenkriegszeit: Otto Neurath, Gerd Arntz*. Vienna: Löcker Verlag.

Stamos, D. N. (2001) quantum indeterminism and evolutionary biology. *Philosophy of Science*, 68(2), 164–84.

Stanley, A. (1995) *Mothers and Daughters of Invention: Notes for a Revised History of Technology*. New Brunswick, NJ: Rutgers University Press.

Steensgaard, N. (1990) The seventeenth century crisis and the unity of Eurasian history. *Modern Asian Studies*, 24(4), 683–97.

Tambiah, S. J. (1990) *Magic, Science, Religion and the Scope of Rationality*. Cambridge: Cambridge University Press.

Temple, R. (1986) *The Genius of China: 3000 Years of Science, Discovery, and Invention*. New York: Simon & Schuster.

Thom, René (1972) *Structural Stability and Morphogenesis* (trans. D. H. Fowler). Reading, MA: W. A. Benjamin (1975).

Thomas, K. (1971) *Religion and the Decline of Magic*. New York: Charles Scribner's Sons.

Thompson, E. (2004) Technology: not artifacts but acts. *American Scientist*, 92(6), 576–7.

Thompson, E. P. (1968) *The Making of the English Working Class*. Harmondsworth: Penguin.

Thompson, W. (1977) *William Morris: Romantic to Revolutionary*. New York: Pantheon Books.

Thornhill, R. (1979) Adaptive female mimicking behavior in a scorpionfly. *Science*, new series, 205(4404), 412–14.

Toulmin, S. (1961) *Foresight and Understanding*. Bloomington: Indiana University Press (selection in Scharff and Dusek, pp. 109–16).

Tuana, N. (1996) Revaluing science: starting from the practices of women. In L. H. Nelson and J. Nelson (eds), *Feminism, Science, and the Philosophy of Science*. Boston: Kluwer Academic (also in Scharff and Dusek, pp. 116–25).

BIBLIOGRAPHY

Turkle, S. (1984) *The Second Self: Computers and the Human Spirit*. New York: Simon & Schuster.

Turkle, S. (1995) *Life on the Screen: Identity in the Age of the Internet*. New York: Simon & Schuster.

Turschwell, P. (2001) *Literature, Technology and Magical Thinking, 1880–1920*. Cambridge: Cambridge University Press.

Uebel, T. E. (ed.) (1991) *Rediscovering the Forgotten Vienna Circle: Austrian Studies on Otto Neurath and the Vienna Circle*. Boston: Kluwer Academic.

Veblen, T. (1899) *The Theory of the Leisure Class*. New York: Mentor (1953).

Veblen, T. (1904) *The Theory of Business Enterprise*. New York: A. M. Kelley, Bookseller.

Veblen, T. (1918) *The Higher Learning in America: A Memorandum on the Conduct of Universities by Business Men*. Stanford, CA: Academic Reprints.

Veblen, T. (1921) *The Engineers and the Price System*. New Brunswick, NJ: Transaction Publishers (1983).

Virilio, P. (2000) *Polar Inertia* (trans. P. Camiller). Thousand Oaks, CA: Sage.

Vogel, H. U. (1993) The Great Well of China. *Scientific American*, June, 86–96.

Vogel, S. (1996) *Against Nature: The Concept of Nature in Critical Theory*. Albany: State University of New York Press.

Vogt, W. (1948) *The Road to Survival*. New York: Sloan.

von Glasersfeld, E. (1995) *Radical Constructivism: A Way of Knowing and Learning*. London: Falmer Press.

Vygotsky, L. S. (1925–34) *Thought and Language*. Cambridge, MA: MIT Press (1986).

Vygotsky, L. S. (1925–34) *Mind in Society: The Development of Higher Psychological Processes*. Cambridge, MA: Harvard University Press (1978).

Wajcman, J. (1991) *Feminism Confronts Technology*. University Park: Pennsylvania State University Press.

Walter, E. V. (1985) Nature on trial: the case of the rooster that laid an egg. In E. V. Walter, V. Kavolis and E. Leites (eds), *Civilizations East and West: A Memorial Volume for Benjamin Nelson*. Atlantic Highlands, NJ: Humanities Press.

Warren, K. (2001) Introduction to Part 3, Ecofeminism. In M. E. Zimmerman, J. B. Callicott, G. Sessions, K. Warren and J. Clark (eds), *Environmental Philosophy: From Animal Rights to Radical Ecology*, 3rd edn. Upper Saddle River, NJ: Prentice Hall.

Wartofsky, M. (1979) *Models: Representation and the Scientific Understanding*. Boston: Reidel.

Watson, J. B. (1925) *Behaviorism*. New York: Norton.

Watson, J. B. (1926) What the nursery has to say about instincts. In C. Murchison (ed.), *Psychologies of 1925*. Worcester, MA: Clark University Press.

Watt, K. E. F. (1968) *Ecology and Resource Management: A Quantitative Approach*. New York: McGraw-Hill.

Weber, M. (1904) *The Protestant Ethic and the Spirit of Capitalism* (trans. T. Parsons). London: Routledge (2001).

BIBLIOGRAPHY

Weber, M. (1914) *Economy and Society: An Outline of Interpretive Sociology* (ed. G. Roth and C. Wittich, trans. E. Fischoff). New York: Bedminster Press (1968).

Weber, M. (1920) *The Rational and Social Foundations of Music* (trans. and ed. D. Martindale, J. Riedel and G. Neuwirth). Carbondale: Southern Illinois University Press (1958).

Weber, M. (1920/1a) *The Religion of China: Confucianism and Taoism* (trans. H. Gerth). Glencoe, IL: Free Press (1951).

Weber, M. (1920/1b) *The Religion of India: The Sociology of Hinduism and Buddhism* (trans. H. Gerth and D. Martindale). Glencoe, IL: Free Press (1958).

Weber, M. (1920/1c) *Ancient Judaism* (trans. H. Gerth and D. Martindale). Glencoe, IL: Free Press (1952).

Webster, C. (1982) *From Paracelsus to Newton: Magic in the Making of Modern Science.* Cambridge: Cambridge University Press.

Weinberg, A. (1966) Can technology replace social engineering? *University of Chicago Magazine*, October. Reprinted in M. Teich (ed.), *Technology and the Future*, 9th edn. New York: St Martin's Press (2003).

Weiner, J. S. (1955) *The Piltdown Forgery*. Oxford: Oxford University Press.

Werskey, G. (1978) *The Visible College: The Collective Biography of British Scientific Socialists of the 1930s*. New York: Holt, Rinehart, and Winston.

Westfall, R. (1980) *Never at Rest: A Biography of Sir Isaac Newton*. Cambridge: Cambridge University Press.

White, L. J. (1962) *Medieval Technology and Social Change*. Oxford: Oxford University Press.

White, L. J. (1978) *Medieval Religion and Technology*. Berkeley: University of California Press.

Whitehead, A. N. (1922) *The Principle of Relativity*. New York: Barnes & Noble (2005).

Whitehead, A. N. (1925) *Science and the Modern World*. New York: Macmillan.

Whitehead, A. N. (1927) *Process and Reality: An Essay on Cosmology* (ed. D. R. Griffin and D. W. Sherburne). New York: Free Press (1978).

Williams, L. P. (1966) *Michael Faraday: A Biography*. New York: Simon and Schuster.

Williams, W. A. (1964) *The Great Evasion: An Essay on the Contemporary Relevance of Karl Marx*. New York: Hill and Wang.

Wilson, B. (ed.) (1970) *Rationality*. Oxford: Blackwell.

Wilson, E. O. (1978) *On Human Nature*. Cambridge, MA: Harvard University Press.

Wilson, E. O. and MacArthur, R. (1967) *Theory of Island Biogeography*. Princeton, NJ: Princeton University Press.

Winner, L. (1977) *Autonomous Technology: Technics-out-of-control as a Theme in Political Thought*. Cambridge, MA: MIT Press (selection in Scharff and Dusek, pp. 606–11).

Winner, L. (1993) Social constructivism: opening the black box and finding it empty. *Science as Culture*, 16, 427–52 (also in Scharff and Dusek, pp. 233–43).

Winograd, T. and Flores, F. (1986) *Understanding Computers and Cognition*. Reading, MA: Addison-Wesley.

BIBLIOGRAPHY

Wittgenstein, L. (1931) *Remarks on Frazer's Golden Bough* (trans. A. C. Miles, ed. and rev. R. Rhees). Doncaster: Brynmill Press (1979).

Wolfe, T. (1979) *The Right Stuff*. New York: Farrar, Straus, Giroux.

Wolin, R. (2001) *Heidegger's Children: Hannah Arendt, Karl Loewith, Hans Jonas, and Herbert Marcuse*. Princeton, NJ: Princeton University Press.

Wolpert, L. (1994) Do we understand development? *Science*, 266, 571–2.

World Commission on Environment and Development (1987) *Our Common Future*. Oxford: Oxford University Press.

Worster, D. (1977) *Nature's Economy: The Roots of Ecology*. Garden City, NY: Doubleday Anchor Press.

Wright, B. D. (ed.) (1987) *Women, Work, and Technology: Transformations*. Ann Arbor: University of Michigan Press.

Wright, R. (1995) The biology of violence. *New Yorker*, March 13, 67–77.

Yates, F. (1964) *Giordano Bruno and the Hermetic Tradition*. New York: Random House.

Yates, F. (1968) The hermetic tradition in Renaissance science. In C. Singleton (ed.), *Art, Science and History in the Renaissance*. Baltimore: Johns Hopkins University Press.

Yates, F. (1972) *The Rosicrucian Englightenment*. Boulder, CO: Shambala.

Young, R. M. (1985) *Darwin's Metaphor: Nature's Place in Victorian Culture*. Cambridge: Cambridge University Press.

Zea, L. (1944) *Positivism in Mexico* (trans. J. H. Schulte). Austin: University of Texas Press (1974).

Zea, L. (1949) Positivism and porphirism in Mexico. In F. S. C. Northrop (ed.), *Ideological Differences and World Order*. New Haven, CT: Yale University Press.

Zeitlin, I. (1968) *Ideology and the Development of Sociological Theory*. New York: Prentice Hall (7th edn 2001).

Zilsel, E. (1942) The sociological roots of science. In D. Raven, W. Krohn and R. S. Cohen (eds), *The Social Origins of Modern Science*. Boston: Kluwer Academic (2000).

Ziman, J. (1968) *Public Knowledge: The Social Dimension of Science*. Cambridge: Cambridge University Press.

Zimmerman, M. E. (1990) *Heidegger's Confrontation with Modernity: Technology, Politics, Art*. Bloomington: Indiana University Press.

Index

INDEX

INDEX

INDEX

INDEX

military industrial complex   49
Mill, John Stuart   72, 91
minitel   103
Minsky, Marvin   80
Modern Times (Chaplin)   131
Mohists   114, 171
Monastery   97
monism   180, 181, 187, 188
Morris, William   2, 181
Morrison, Philip   153
Moss, Lenny   132
Mughal   173
multiples   95
Mumford, Lewis   4, 31, 32, 102, 113,
   116, 118, 119, 122, 123–6, 129,
   132, 134, 138, 184
Murray, Gilbert   163
Museum of Modern Art   33
Muslim   20
mythic thought   164–5

Nabi, Isador   185
Naess, Arne   37, 187, 190
Nagel, Thomas   137
Native American   173
natural gas wells   170
nature philosophers   178, 179
Nazis   18, 20, 108, 109, 128, 165, 183,
   191
Needham, Joseph   114, 169, 170–1,
   174
negative theology   30
Nelson-Burns, Leslie   136
neo-Kantianism   74, 200
Neolithic hunter-gatherer   126
neo-Malthusianism   192, 194
neo-Marxism   4
neo-Platonism   40
neural networks   80
Neurath, Otto   9
New Age   88, 153–4, 163, 165, 181,
   187

New Atlantis   41, 43
New Eloise, The   177
New Organon   42
new priesthood   125
Newton, Isaac   14, 34, 44, 87, 88, 95,
   142, 145–6, 162, 170, 172, 178,
   179
Nietzsche, Friedrich   126, 129, 143
Noble, David   102, 147, 206
nominalism   27, 28

object relations theory   147
Ockham, William of   28
Odum, H. T.   185
Oersted, Hans Christian   179
Ogburn, William F.   94
Oldenburg, Henry   146
Oppenheimer, J. Robert   49, 105, 109
oracles   167
oral communication   96, 97, 100, 157
organicism   180, 181
organismic biologist   180
Ortega y Gasset, José   112
orthodox Marxism   176
Oscar II of Sweden and Norway   87
Ottoman Empire   173
ovens   139

Pacey, Arnold   35, 103, 161, 168
Paijing   116
panpsychism   188, 209
Papert, Seymour   80
Paracelsus, Theophrastus
   Bombastus   162
paradigm   13–14, 22, 24, 53, 55
Parsons, Talcott   186
Pascal, Blaise   78
Pasteur, Louis   35
Pavlov, Ivan   47
Peirce, Charles S.   81, 208
Penfield, Wilder   91
penicillin   34

240

INDEX

242

INDEX

sustainability 194–6, 197
symbolic forms (Cassirer) 164
symbolic logic 81, 121, 162
synthesis 59, 200
Syracuse 115

Tansley, A. G. 184–5
Taoists 114
Taylorism 123
technocracy 2, 17, 38–52, 54, 60, 61,
    66, 96, 125, 133, 146, 176
technocrats 53
technological determinism 3, 84, 86
technological system 33, 35–6, 63, 90,
    107–8, 109, 110, 138, 148–9, 168,
    204, 208
technology as rules 31, 32, 106
technoscience 23
technostructure 50–1, 52
Teller, Edward 49, 109
things in themselves 8, 199, 200
Thurber, James 16, 136
thyroid hormone 169
Timaeus 113
tool-makers 113, 118–19, 134
tool-making 112–13, 117–18, 120,
    126–7, 133–4, 135
tools 12, 13, 31, 32, 33, 35, 36, 37,
    113, 114, 117–18, 120, 125, 127,
    134, 135, 164, 168
Toulmin, Stephen 62, 74, 137
transcendentality 2, 68–9, 178
trial and error 33, 34
Turkle, Sherry 128
two standpoints (Kant) 92
Tycho Brahe 171
typewriter 144

ultrasound 143–4
uncertainty principle 88
underdetermination 16–17
unintended consequences 104, 106

United Nations Commissions on
    Sustainable Development 196
Upjohn Corporation 141
USSR 18, 44, 48, 49, 51, 95, 110, 123,
    125
utilitarianism 56, 63, 65

valuing life 64
Veblen, Thorstein 39, 46, 47
verification theory 7
Vico, Giambattista 108, 199
Vienna Circle 7, 13, 82
Vietnam War 49, 107
Virginia Company 142
Virilio, Paul 5, 209
von Neumann, John 49
vulcanization 34
Vygotsky, Lev 202

Wallace, Alfred Russel 95, 148
Watson, John B. 47, 86
Watt, Kenneth J. 185
Weber, Max 32, 53–4, 57, 60, 106,
    116, 198
Weinberg, Alvin 125
Weiss, Paul 180
Wells, H. G. 185
Wheeler, John 188
Wheeler, William Morton 186
Whiston, William 162
Whitehead, Alfred North 5, 38, 72,
    119, 178, 186, 208–9, 210
Whitney, Eli 139
Wiener, Norbert 82
Wilson, Edward O. 121, 185
Wilson, Harold 48
Windelband, Wilhelm 74
Winograd, Terry 82
Wittgenstein, Ludwig 29, 78, 163
Wizards of Armageddon (Kaplan) 125,
    166
Wolfe, Tom 154

243

INDEX